CENTRIFUGAL PUMPS

CENTRIFUGAL PUMPS

A GUIDE FOR PUMP USERS

K M Srinivasan
BE (Hons) (Mech), PhD (USSR)
Former Dean (R&D)
Mechanical Sciences, Kumara Guru College of Technology
Ganapathy, Coimbatore
Tamil Nadu

Publishing Globally
NEW AGE INTERNATIONAL (P) LIMITED, PUBLISHERS
New Delhi • Bangalore • Chennai • Cochin • Guwahati
Hyderabad • Kolkata • Lucknow • Mumbai
Visit us at **www.newagepublishers.com**

Published by New Age International (P) Ltd., Publishers
First Edition: 2016

BRANCHES

- **Bangalore** 37/10, 8th Cross (Near Hanuman Temple), Azad Nagar, Chamarajpet Bangalore-560 018, Tel.: (080) 26756823, Telefax: 26756820 **E-mail: bangalore@newagepublishers.com**
- **Chennai** 26, Damodaran Street, T. Nagar, Chennai-600 017, Tel.: (044) 24353401 Telefax: 24351463, **E-mail: chennai@newagepublishers.com**
- **Cochin** CC-39/1016, Carrier Station Road, Ernakulam South, Cochin-682 016 Tel.: (0484) 2377303, Telefax: 4051303 **E-mail: cochin@newagepublishers.com**
- **Guwahati** Hemsen Complex, Mohd. Shah Road, Paltan Bazar, Near Starline Hotel Guwahati-781 008 Tel.: (0361) 2513881, Telefax: 2543669 **E-mail: guwahati@newagepublishers.com**
- **Hyderabad** 105, 1st Floor, Madhiray Kaveri Tower, 3-2-19, Azam Jahi Road, Near Kumar Theater Nimboliadda, Kachiguda, Hyderabad-500 027, Tel.: (040) 24652456 Telefax: 24652457 **E-mail: hyderabad@newagepublishers.com**
- **Kolkata** RDB Chambers (Formerly Lotus Cinema) 106A, 1st Floor, S N Banerjee Road Kolkata-700 014 Tel.: (033) 22273773, Telefax: 22275247 **E-mail: kolkata@newagepublishers.com**
- **Lucknow** 16-A, Jopling Road, Lucknow-226 001, Tel.: (0522) 2209578, 4045297 Telefax: 2204098, **E-mail:lucknow@newagepublishers.com**
- **Mumbai** 142C, Victor House, Ground Floor, N.M. Joshi Marg, Lower Parel Mumbai-400 013, Tel.: (022) 24927869, Telefax: 24915415 **E-mail: mumbai@newagepublishers.com**
- **New Delhi** 22, Golden House, Daryaganj, New Delhi-110 002 Tel.: (011) 23262368, 23262370, Telefax: 43551305 **E-mail: sales@newagepublishers.com**

ISBN: 978-81-224-3883-3

₹ 499.00

C-15-05-8519

Printed in India at Glorious Printers, New Delhi.
Typeset at In-house, New Delhi.

PUBLISHING GLOBALLY
NEW AGE INTERNATIONAL (P) LIMITED, PUBLISHERS
7/30 A, Daryaganj, New Delhi - 110002
Visit us at **www.newagepublishers.com**

PREFACE

It is my intention, to bring out a book on using pumps in proper manner and to give information, useful to the pump purchasers and customers. Pump is supposed to be the HEART of industries. Pumps are used, not only in domestic and agricultural areas, but also in industries. Necessary and essential materials, for the selection of pumps, based on the experience gained by different authors around the world, are also given in a nutshell. Quick reference can be made by referring the monograms and charts, given.

I will be very much grateful for the valuable assistance given by my wife Smt. Nalini Srinivasan, and my two sons, Sri Muthuraman Srinivasan and Sri Jagan Mohan Srinivasan, in producing this manuscript. Also I am very much thankful to M/s New Age International Publishers, for publishing this book in the correct and commendable form. It is very helpful and useful, if readers go through this book, thoroughly and give their useful suggestions and their opinions to improve the quality of this book.

Dr. K.M. Srinivasan

CONTENTS

1 FUNDAMENTALS

1.1 INTRODUCTION

A pump is a mechanical device, which converts mechanical energy into hydraulic energy, which consists of pressure, velocity, static height either individually or the combination of any two of the three or all the three, put together. The energy supplied to the fluid can be used for different purposes such as (1) to raise the fluid to a higher level (domestic use) (2) to overcome the frictional losses in a pipeline (chemical industries, water supply distribution in cities and in agriculture) (3) to do external work through a hydraulic device (industries) or (4) all combined.

The main classification of pumps is as follows:

(*a*) Non-positive displacement or roto dynamic pumps.

(*b*) Positive displacement pumps.

1.1.1 Roto Dynamic Pumps

Radial, Mixed or Diagonal and propellor or Axial flow centrifugal pumps belong to this category.

1.1.1.1 Radial Flow Centrifugal Pumps

A centrifugal pump gets its name from the term centrifugal force defined as the force exerted by a body moving in a circular path. This centrifugal force is used in pumps to transfer the liquid, from one place to another.

The principle of operation can best be explained by a simple example. Consider a boy holding a bucket with water, as in Fig. 1.1.

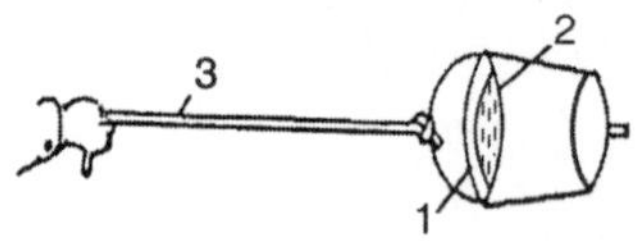

(1) Bucket, (2) Water and (3) Rope

Fig. 1.1: Principle of operation of centrifugal pump.

The boy whirls the bucket with water around himself. Not even a single drop of water spills down, so long, the boy rotates the bucket faster enough. The force holding the water from spilling is known as the centrifugal force. Now suppose a hole is drilled at the bottom of the bucket. A stream of water will then rush out from the bucket through this hole, due to the centrifugal force, created when the bucket is rotated. This stream would be continues, if the boy's body is considered as the suction pipe, through which water will be coming continually, then through his arm, and finally to the bucket. The boy's arm is the impeller; the bucket is the casing throwing a stream of water out of the pump. The casing has been introduced to guide water effectively to the discharge pipe, which is the hole in the bucket. The liquid forcing out creates a vacuum in the arm and hence continuous flow is maintained by the liquid forcing itself into the casing by outside atmospheric pressure.

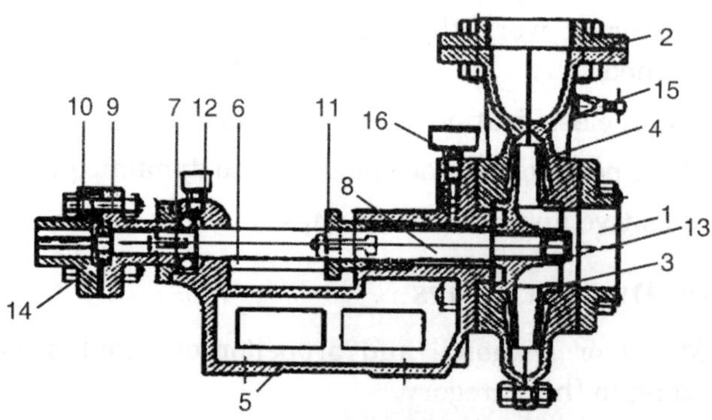

1. Suction Flange
2. Delivery Flange
3. Impeller
4. Volute Casings
5. Bearing Bed
6. Shaft
7. Deep Groove Ball Bearing
8. Bush
9. Flexible Coupling (Pump Side)
10. Flexible Coupling (Motor Side)
11. Gland
12. Bearing Cap
13. Impeller Nut
14. Coupling Nut
15. Air Cock
16. Grease Cup

Fig. 1.2(*a*): Single bearing supported single stage centrifugal pump.

A single stage centrifugal pump, shown in Fig. 1.2 (*a* & *b*) can develop only upto a head of about 30 m (100 feet). For higher heads, we must go for multistage pumps. There are two types of multistage pumps.

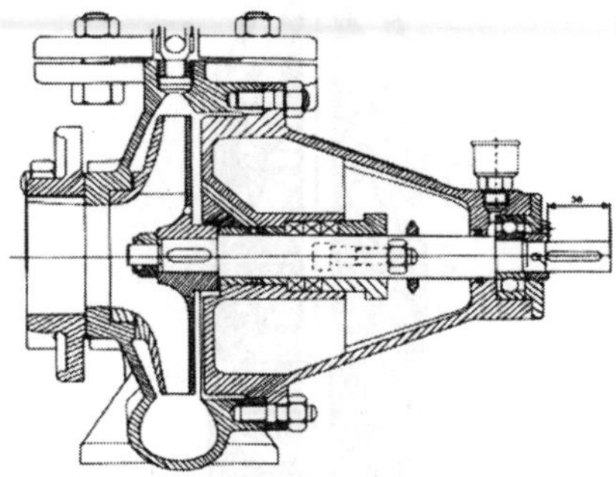

Fig. 1.2(*b*): Single stage backpull out type centrifugal pump.

1. If the pump impellers are connected in series, *i.e.*, one after another, pumping head is the sum of the total head given by the individual impellers. In that case, same quantity of water passes through all these impellers. One such pump is shown in Fig. 1.3. Deep well pumps and submersible (Fig. 1.4*a* & *b*) pumps are the examples of pumps, where impellers are connected in series. These pumps have their axis vertical and are used mainly for bore wells. Figure 1.4 shows the schematic arrangement of these pumps.

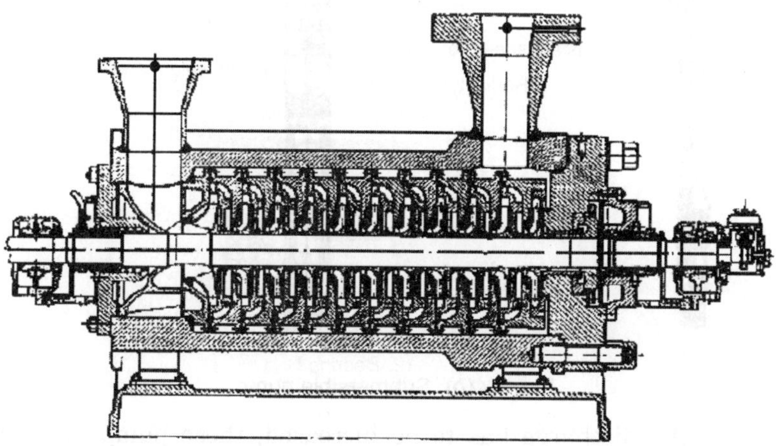

Fig. 1.3: A 10 stage (10 Impellers) multistage pump.

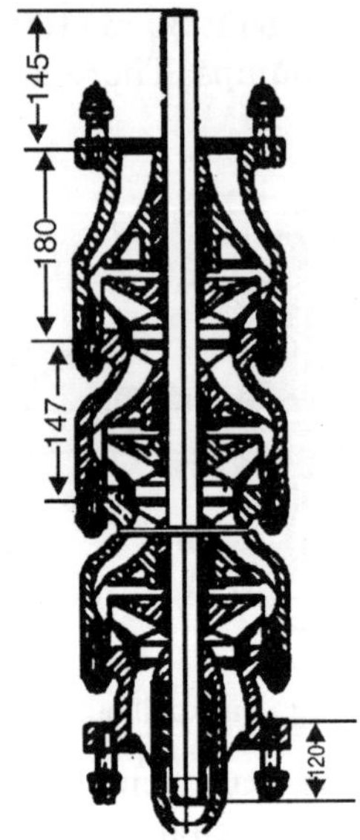

Fig. 1.4(*a*): Deep well turbine pump.

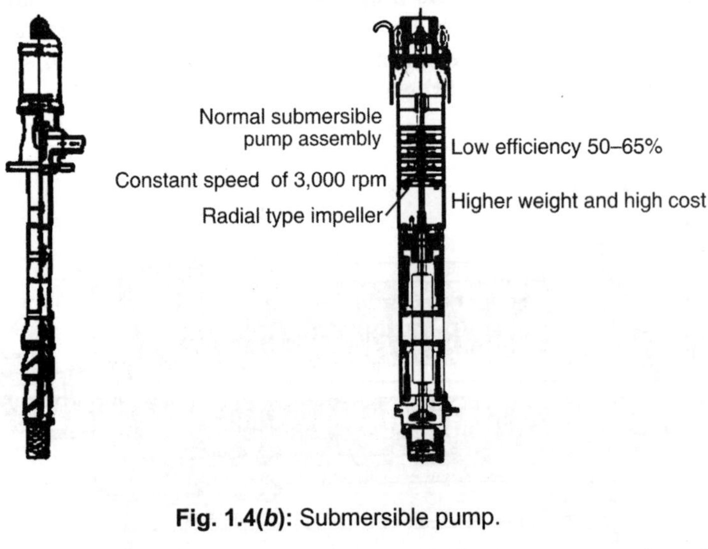

Fig. 1.4(*b*): Submersible pump.

2. If the discharge has to be increased, these pump impellers must be connected in parallel. In this case the total head will remain same as individual total head. Discharge will be multiplied. A double suction pump is an example of parallel operation of two impellers and is shown in Fig. 1.5.

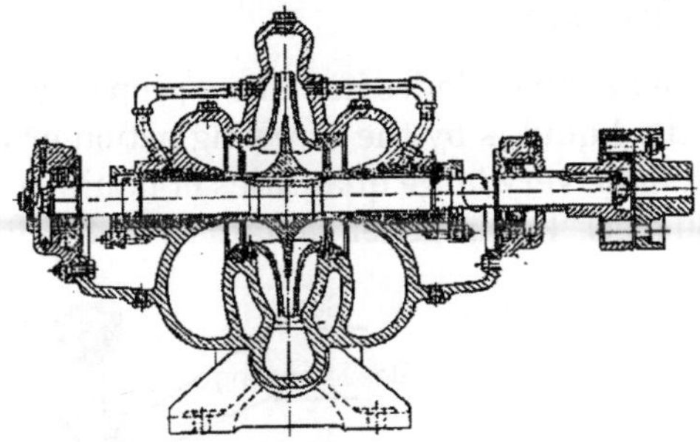

Fig. 1.5: A double suction pump.

Centrifugal pump is the most efficient, for handling small quantities of liquid at high and medium heads. It can handle liquids with foreign materials, such as sand, mud, drainage water also. They are used for several types of pumping services such as (*a*) general water supply e.g. Domestic and industrial water supply etc., (*b*) pressure service e.g. boiler feeding, hydraulic press, multi storied flats, etc., (*c*) circulating service e.g. refrigerators, air conditioners, municipal water supply, etc., (*d*) transporting service e.g. loading and unloading of tanks and (*e*) process service e.g. pumping of chemicals, paper stock, acids, alkalies and neutral solutions like milk, distilled water etc.

The main features of a centrifugal pump are (*a*) simple construction, (*b*) no close clearances, (*c*) small floor space, (*d*) low initial cost, (*e*) easy maintenance, (*f*) continuous flow, (*g*) quiet in operation, (*h*) no reciprocating parts and (*i*) high speed of rotation.

1.1.1.2 Mixed Flow Pump

The mixed flow pump has an impeller which imparts the energy to the fluid by both centrifugal and axial forces, when its impeller rotates inside its casing. Figure 1.6 gives a mixed flow pump. They are used mainly for medium heads and medium discharges.

Fig. 1.6: Mixed flow pump.

1.1.1.3 Axial Flow Pump

These pumps have impeller blades just like a propeller (Fig. 1.7). All the energy given to the liquid is by the screwing action or axial force only. This pump can handle very large quantities of liquid at very low heads, 1 to 10 m (3 to 30 feet). The impeller contains 3 to 4 number of blades.

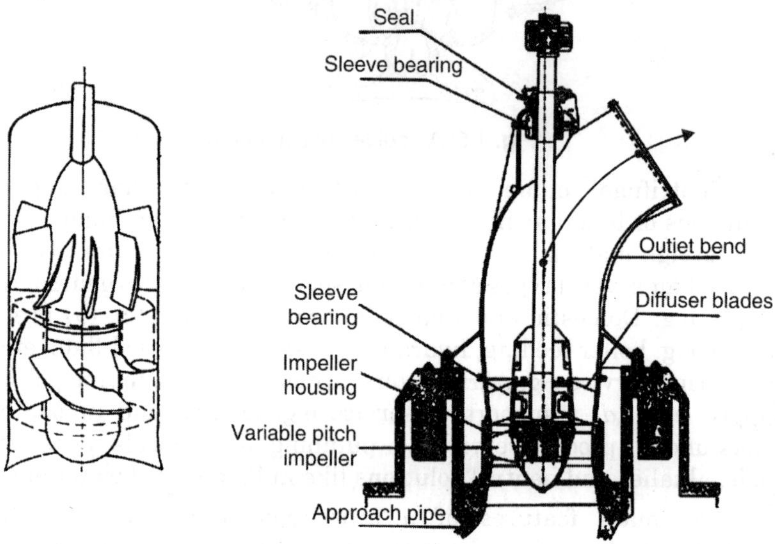

Fig. 1.7: Axial flow pump.

Apart from these general types, there are many other types with their own advantages, disadvantages and fields of operation. Liquids with solid particles such as drainage, molasses, crude oil etc. can also be handled, by these pumps. A careful study of all the types of pumps, their principles, characteristics and application are essential before suggesting a correct pump for the job.

1.1.2 Positive Displacement Pumps

Positive displacement pumps are of two types: 1. Reciprocating pumps and 2. Rotary pumps. Figure 1.8 shows the line diagram of a reciprocating pump. As the crank rotates, the piston moves to and fro. During its motion, it creates suction and pressure alternatively. During suction stroke, the cylinder is filled with liquid and during pressure stroke it is pumped through the delivery.

These pumps are suitable for high delivery pressures, and low capacities. They are used mainly for feeding small boilers, gasoline pumping, fuel oil pumping and small sprinkler or drip irrigation, etc.

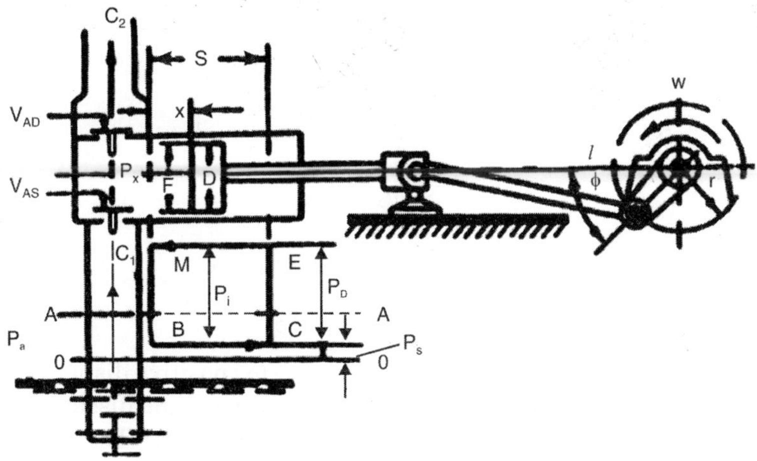

Fig. 1.8: Reciprocating pump.

The main features of a reciprocating pump are (*a*) positive displacement, (*b*) slow-speed (50–100 rpm), (*c*) self-priming, (*d*) high starting torque and (*e*) same discharge for different heads. Head and Discharge are independent.

In a rotary pump, the liquid is trapped between blades and pushed around a closed casing, thus imparting energy to the liquid. A few types of rotary pumps used are (*a*) Vane pump, (*b*) Lobe pump and (*c*) Gear pump. Simple sketches of these pumps are given in Figs. 1.9, 1.10 and 1.11.

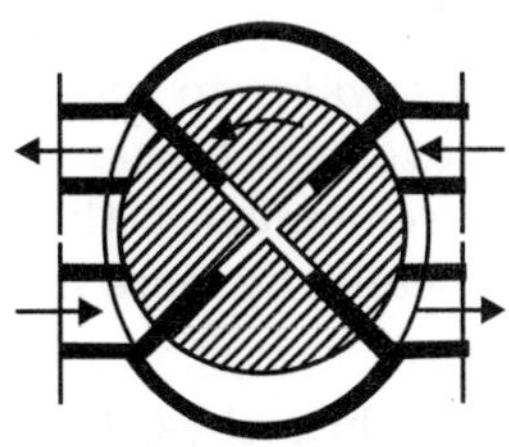

Fig. 1.9: Vane pump.

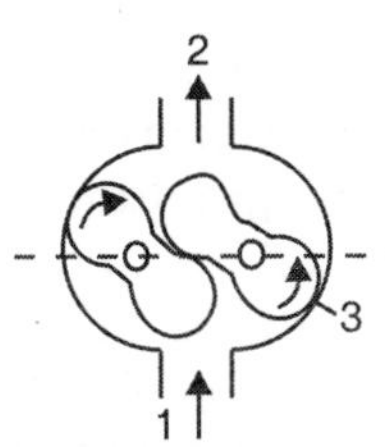

(1) Suction, (2) Delivery and (3) Lobe

Fig. 1.10: Lobe pump.

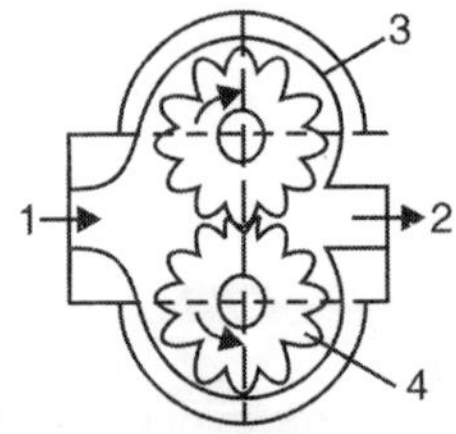

(1) Suction, (2) Delivery and (3) Gears

Fig. 1.11: Gear pump.

The main features of a rotary pump are (*a*) positive displacement, (*b*) medium speed, (*c*) self-priming, (*d*) fairly constant discharge and (*e*) medium heads, but very low suction lift (upto 0.5 m). They are used mainly for pumping lubricating oils, fuel oils, etc., and for loading and unloading of oil tanks, vessels, etc.

1.2 A FEW POINTS ABOUT PUMPS

1.2.1 Atmospheric Pressure, Effect of Earth's Gravity and Suction Effect on Pumps

You are perhaps one of those, who want to purchase a pump. If so, it is necessary, that you understand a few points.

Do you know what you do, when you take water from a well by a pump? In order to move the water away from the well, you must add energy to water. For this purpose, you use a pump. The pump supplies energy to water and makes water to move upwards from the well. During the process of its movement upward, the energy it possesses is used. The point, you should remember, is that, with a centrifugal pump you cannot lift water, practically, by more than 5 to 6 m (15 to 20 feet). This needs some explanation.

Fruits are held in tree branches, by the force exerted by the branches on the fruit. When the fruits separate from the tree branches, the fruits fall on the ground. This is because earth has gravitational force and it draws the fruit towards its centre. As long as the fruit is in the tree, it is held by the force exerted by the branch, in a direction opposite to gravity. This same gravitational force of the earth holds the atmospheric air, which roughly extends to a height of about 10336 metres above the earth. If there is no gravity, the atmospheric air would have been lost in space. Any object, including air, has weight. This weight exists because the gravitational force of the earth pulls the object towards the earth. A column of air, which has an area of one square metre at its base and stands vertically to a height of 10336 metres weighs 10336 kg. We really carry this weight on our shoulders all the time. This pressure of 10336 kg/m^2 or 1.0336 kg/cm^2 is called the atmospheric pressure of air.

One other point is that the Mercury is a very heavy liquid. If we pour mercury into a glass tube and put tube in an inverted position in such a way that its open end of the mouth is immersed inside the pool of mercury and its closed end is at the top, the column of mercury in the inverted tube stands to a height of only 76 cms and the remaining space above the mercury level is empty. But if the tube is less than 76 cms in height then the entire tube length remains filled with mercury. This is because the pressure of the atmosphere exerts an upward force from the bottom of the tube. This pressure can keep the mercury to a height of 76 cms (about 30 inches). If we use water instead of mercury in the tube, the water will stand to a height of 10.336 metres (34 feet).

We may now imagine a smooth tube of several metres long, with a piston into it. Let the tube be placed vertically in a large vessel of water. Let us draw the piston upward. The water in the vessel rises inside the tube. As we draw the piston upward water goes on rising. Water will rise to a height (called static suction height, H_s) of 10.336 metres (34 feet) only. This is the maximum height of a water column, which the atmospheric pressure can support. This is how, a pump also works (Fig. 1.12).

Let us study, the diagram of a centrifugal pump and its parts (Figs.1.1 and 1.2). Any centrifugal pump mainly consists of impeller, casing, bearing bracket, and a shaft, on which impeller at one end and a coupling on the other end are mounted. The arrangement consisting of a number of vanes with the "shroud" on either side is called the "impeller". Impeller is inside a casing and the casing is connected to suction and a delivery pipe. A "Foot valve" is fitted to the end of the suction pipe which is immersed in water. This foot valve allows water in one direction only. It will allow water to get into the suction pipe, but will not allow water from the suction pipe to come out.

Before a pump is started, the suction pipe and the casing of the impeller are all filled with water. This is called "Priming" the pump. After priming, the pump is started. The impeller rotates inside the casing. The vanes create a circular motion in water within the casing. Scientists tell that, when any object is subjected to a force in radially outward direction from the centre, this force is called "centrifugal force" and this force moves the liquid away from the centre of rotation. For example, when you travel in a car or bus and when the car takes a turn towards your right, your body is thrown towards your left. This is due to centrifugal force. When the impeller of the centrifugal pump goes on rotating, the primed water is first thrown out and driven to the periphery of the casing. This water gets collected and flows out of the delivery pipe attached to the casing. With its exit, a vacuum is created. Immediately the atmospheric pressure starts acting. It presses the water in the well. Water rises in the suction pipe. This is how water is lifted from the well. This water is driven into the suction pipe through the foot valve and strainer. The water then enters the impeller. The rotating impeller again acts on the water and drives it to the casing and then to the delivery pipe. This goes on as long as the pump works.

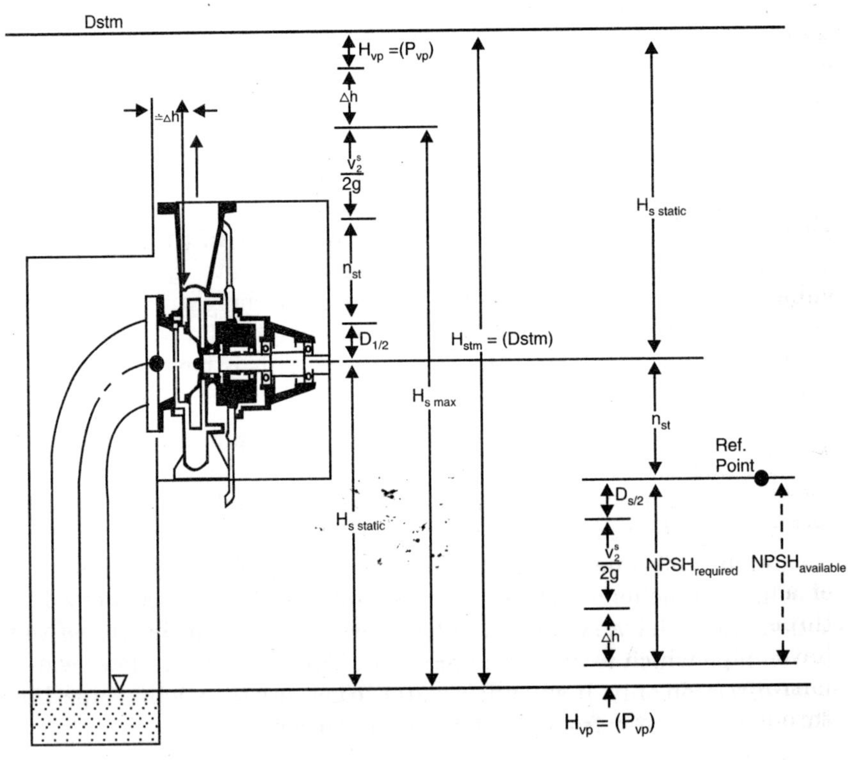

Fig. 1.12: Suction height for a pump.

It is commonly said that a centrifugal pump sucks water from a well. The centrifugal pump does not really suck water. It merely creates a vacuum in the suction pipe, when the impeller rotates inside the casing. In actual practice, it is found that a centrifugal pump installed near sea level can lift water upto a height of about 10.336 metres (34 feet) only. Water does not rise above 10.336 metres (34 feet).

Atmospheric pressure presses the water to move inside the suction pipe of the pump, theoretically to a height of 10.336 metres (34 feet). At that height, the weight of this column of water exercises a downward pressure. This downward pressure is equal to the atmospheric pressure, acting upwards. Equilibrium is established.

We saw, how the atmospheric pressure can support a column of 10.336 metres of water and how a centrifugal pump installed at mean sea level can lift water, theoretically, upto a height of 10.336 metres, but practically only upto 5 to 6 m. These facts, ascertainable from laboratory experiments, may now be applied to the field conditions under which a pump is installed.

First, a pump installed at sea level, can theoretically lift water upto a height of 10.336 metres. If we move away from the sea level and go to a mountain, say Ooty, which is about 2,500 metres above mean sea level, the atmospheric pressure will be less. Scientists have found out that, for an elevation of every 900 metres above mean sea level, there will be a decrease in the atmospheric pressure by 1 metre of water. If your well is located at Ooty, the atmospheric pressure will be 10.336 – (2500/900) = 10.336 – 2.78 = 7.556 m only.

Second, when water runs through a pipe, the running water is subjected to friction, due to the flow of water over the rough inside surface of the pipe line. This friction interferes with the free flow of water. As a result, a pump installed at mean sea level will not have a suction lift of exactly 10.336 metres. It will be less due to the losses in the pipe. This loss is called **"loss of head due to friction"**. If the pipe, through which water flows, is very long, the total amount of friction is also large and therefore loss of head is also proportionately more. This loss is called **Major Losses**.

Third, water, pumped out, may be allowed to flow through a pipe of a large or small diameter. For instance, when several cars pass through a narrow road, there is congestion and bottleneck and cars have to proceed slowly. Similarly when water flows through a pipe of narrow diameter, there will be more friction than when the water flows through a pipe of a larger diameter.

Fourth, bends tee joints, valves, and change in the pipe size, in the pipe lines also create friction. There is loss of head on this account also. These losses are called **Minor Losses**.

When water is pumped out of a well, the following are the sources of loss of head:

(*a*) The foot valve and the strainer at the bottom of the suction pipe offers friction, because the one way valve in the foot valve has to be opened by the force of the incoming water only.

(*b*) As the water passes through the suction pipe, there will be resistance from the wall of the pipe, called friction.

(*c*) Then, if there are bends or valves etc., they also cause friction.

For example, if we assume that, the loss of head in case of the foot valve is (¼) m, due to friction in the suction pipe is 2 m and another (¾) m in bends, there is a total loss of 3 metres. Suppose a well, installed at mean sea level, has a depth of 8 metres between the pump impeller and water level in the well, the water will not rise to the impeller because 3 metres is already lost due to friction at various points, and what is left is only 7.336 metres (atmospheric pressure at mean sea level 10.336 m – friction 3 m = 7.336 m). This suction head is not enough for the water in the well to reach the impeller which is 8 metres away. So the pump will not work.

In another words, the distance between the pump impeller and water level in the well should be much less than the atmospheric pressure of the place, in order to provide for all losses. An allowance of say 2.5 to 3.5 metres should be made. Thus in a place 900 metres above mean sea level the atmospheric pressure is 10.336 – 1 = 9.336 m. From this, another 3 m may be deducted for frictional losses. The pump impeller should therefore be about 5.5 to 6 m, from the water level in the well.

The following precautions must be taken in installing pumps:

1. Erect the pump as near as possible to the level of the water in the well, so that a minimum suction height is provided.
2. See that the foot valve in the suction pipe is well immersed in the well water, in order to ensure that air particles do not enter the pipe. Normally, foot valve is kept at a minimum depth of 10 times the suction pipe diameter, from the free surface of water. Higher depths are preferable.
3. A perfect vacuum is necessary inside the pump, in order to work efficiently. Therefore all joints should be air tight. Suitable washers must be provided between pipes and all joints, for air tightness.
4. A strainer is situated below the foot valve, at the bottom most point of the suction pipe line. See that the distance between the strainer and the bottom of the suction well or tank floor, and also around the strainer is more than 5 to 6 times the diameter of the suction pipe. This will facilitate free flow of water, even if the meshes in the strainer get clogged.

5. Suction pipe from the pump should be laid at least for 4 to 5 times the diameter of suction pipe horizontally or with a uniform upward slope from the bend, which is connected to the vertical suction pipe leading to the well. This will not allow formation of air pockets, sticking inside the suction pipe, which will reduce the discharge of water from the pump. On no account should the level of the pipe line rise above the pump centre line.
6. Try to use suction pipes of larger diameter. This will minimize friction losses.
7. Avoid a circuitous and long path for the pipe lines. This will also minimize friction losses.
8. Avoid too many bends, valves and fittings. Partially closed valves will also produce more frictional losses. If you have to use bends, avoid sharp bends (called Elbows), use long bends, which reduce losses.
9. Do not allow water to be discharged very near the suction pipe. This may allow air to enter the suction pipe and make the pump inefficient.
10. Try to go in for pumps fitted with medium speed motors in order to have more suction lift.

1.3 HOW TO MEASURE THE QUANTITY OF FLOW IN PUMP AT FIELD?

When the flow rate of the pump, purchased, has to be determined in the field, the maping method is used. No tank, or stop watch, or weighing balance or flow metre is necessary, for flow measurement. The pump is mounted with suction and delivery pipes, near the water source. Care should be taken so that the delivery of water from the delivery pipe is strictly horizontal direction. Allow water from the delivery pipe to fall on the ground. Measure the horizontal distance between the delivery outlet and the point on the ground, where the delivered water touches the ground [distance (*AB*) *X* in Fig. 1.13] and the vertical distance from the water delivery point from the delivery pipe to the ground [distance (*AC*) *Y* in Fig. 1.13]. The following formulae are used to determine the flow rate from the pump.

$$v = \frac{X}{t} \text{ or } t = \frac{X}{v} \qquad ...(1.1)$$

$$Y = \frac{1}{2}\, g\, t^2 \text{ or } \frac{1}{2} g \frac{X^2}{v^2} \qquad ...(1.2)$$

$$v = \sqrt{\frac{gX^2}{2Y}} = 2.215\frac{X}{\sqrt{Y}} \qquad ...(1.3)$$

where v is the velocity in m/sec.

X is the horizontal distance of fall of water from the delivery pipe in metres (m).

Y is the vertical distance between the pipe and the ground in metres (m).

Table 1.1: Velocity of water coming out of the delivery pipe, (v), in m/sec

Y \ X	5 m	6 m	7 m	8 m	9 m	10 m	11 m	12 m	13 m
2 ½ m	7	8.41	9.81	11.21	12.61	14.00	15.41	16.81	18.21
3 m	6.39	7.87	8.95	10.23	11.51	12.70	14.07	15.35	16.62
3 ½ m	5.92	7.10	8.29	9.47	10.66	11.84	13.02	14.21	15.39

$A = \frac{\pi d^2}{4}$ where, d is the diameter in metres. A is the area in sq. metres (= A in sq inches × (0.00064516).

Table 1.2: Area of pipe for the given diameter

(*d*) diameter in inches	**(*A*) Area in Sq. inches**	**(*A*) Area in Sq. Meters**
1 ½ inch	1.7672 sq. inch	1.14×10^{-3}
2 inch	3.1415 sq. inch	2.02×10^{-3}
2 ½ inch	4.9087 sq. inch	3.17×10^{-3}
3 inch	7.0686 sq. inch	4.56×10^{-3}
4 inch	12.5664 sq. inch	8.11×10^{-3}
6 inch	28.2743 sq. inch	18.20×10^{-3}

Q (lit/sec) = A (sq. metres) × v (m/sec) × 1000

Table 1.3: Quantity of flow through the installed delivery pipe, (Q), in lit/sec

(v) m/sec / (d) inch	4 m/sec	5 m/sec	6 m/sec	7 m/sec	8 m/sec	9 m/sec	10 m/sec
1 ½	4.56	5.70	6.86	7.98	9.12	10.26	11.40
2	8.08	10.10	12.12	14.14	16.16	18.18	20.20
2 ½	12.67	15.14	19.00	22.10	25.34	28.50	31.67
3	18.24	22.80	27.36	31.92	36.48	41.04	45.60
4	32.43	40.54	48.65	56.76	64.16	72.97	81.08
6	72.80	91.00	109.20	127.40	145.60	163.80	182.00

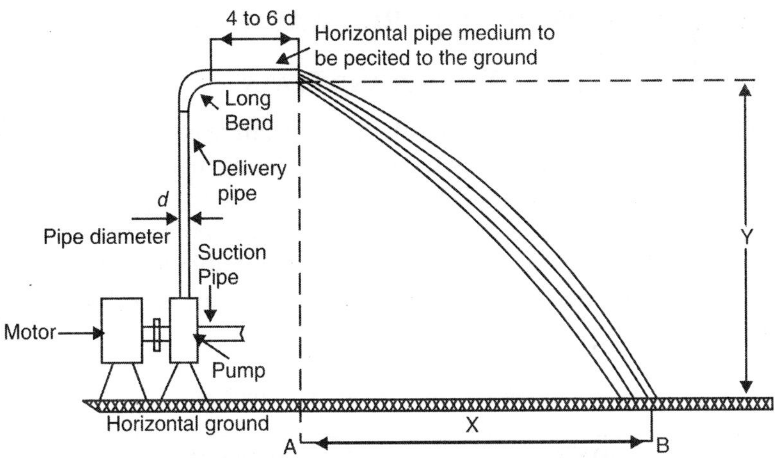

Fig. 1.13: Flow measurement in field.

2 HOW A PUMP IS SELECTED

2.1 CALCULATION OF REQUIRED WATER QUANTITY

2.1.1 For Individual Houses and Flats

House construction programmes both in urban areas, rural areas and in the suburbs of towns are in progress. Many persons look out for installing pump sets in their houses in order that they may live more comfortably. In doing so, they may find it necessary to calculate, what quantity of water they would require and what should be the capacity of the pump set, they need. The following schedule shows the normal requirements of water per day for houses.

Table 2.1: Daily water consumption

Details	Approximate requirement per day
For every member of a family (total requirement including kitchen, bath, laundry, etc.)	135 to 175 litres
For every yielding cow If water is supplied by means of vats or other receptacles for holding water	115 litres
If vats are not used	55 litres
For a bullock or non-yielding cow—If water is fed by means of vats or other receptacles for holding water	65 litres
If vats are not used	36 litres
For a horse	36 litres
For a sheep	5 litres

For every 100 poultry birds	13 litres
For supply water to lawn through a 12 mm (½″) hose with nozzle	750 litres

The above data is based on the wide experience, which many persons have acquired. You will also see that very large quantities of water have been provided for yielding cows. Milk contains as much as 85 percent of water. Heavy yielding cows would require relatively larger quantities of water. You will also see that, where vats or other receptacles are used for providing a ready supply of water, larger requirements of water is estimated. Many authorities say that cows which get a ready supply of water all the time yield more milk. Similarly large supply of water to poultry is recommended because an egg contains as much as 65 % of water and good egg laying birds require larger amounts of water. From the schedule given above, you can calculate, what quantity of water your house will require.

Let us now work out the water requirements for a family of 4 adult members and 2 children. Let us also assume that the family maintains milk yielding cow and a bullock and in addition has 50 poultry birds. The water requirement of this family per day will be as below:

Table 2.2

4 adult members @ 150 litres each	600 litres
2 children, equivalent to one adult	150 litres
One yielding cow (vats are used)	115 litres
One bullock (vats are used)	70 litres
50 poultry birds at 14 litres for every 100 birds	7 litres
Total:	**942 litres**

The smallest size of domestic pump with one inch diameter of suction pipe and one inch diameter of delivery pipe will give a discharge of about 1350 litres per hour. Such a pump is more than enough for this family. We shall deal with water requirements for irrigating crops.

2.1.2 For Agricultural Purpose

Let us now consider how we can calculate water requirements for agricultural purposes. For this purpose, we should understand how the volume of water supplied to a field is calculated.

Suppose we supply water to one Hector (1 H = 2.47 Acres = 10,000 Square Metres) of a field and the height to which the water is allowed to stand in the field is 35 mm. Water, standing 35 mm high in the area of one hector will measure 10,000 × 0.035 = 350 cu.m. Allowing for evaporation losses, seepage etc., it may be necessary to supply about 400 cu.m of water.

Suppose this volume or this land has to be irrigated, in ONE day, then the volume required per day will be (400/1) = 400 cu.m/day. Suppose the pump, you are going to purchase has to work 10 hours per day, then pump must supply (400/10) = 40 cu.m per hour. 1 cu.m = 1000 litres and 1 hour = 3600 seconds. Therefore 40 cu.m/hour = (40 × 1000/3600) = 11.1 litres/sec. You must select a pump giving a discharge (or flow rate) of 11.1 litres per second (or 40 cu.m/hour).

2.1.3 Each Crop Requires Different Quantity of Water

The following table gives the irrigation interval and the water depth required for each crop. The interval of time allowed between two successive irrigations also differs.

Table 2.3: Water required for irrigation

Crop	Irrigation interval in days	Water depth required	
Sugarcane	10–12	20 cm	3.3
Paddy	10–12	23 cm	4.3
Ragi or bajra	25–30	18 cm	3
Cholam or jowar	20	18 cm	3
Cotton	20	18 cm	3
Wheat	40	9 cm	1.5
Bengal gram	40	14 cm	2.2
Groundnut	35	10 cm	1.7
Banana	10	18 cm	2.5
Leafy vegetables	7–10	13 cm	2.0
Onions	9	14 cm	2.2
Tomato	11–12	16 cm	2.8
Potato	11–12	16 cm	2.8
Other vegetables	10–12	16 cm	2.8
Carrots	8–10	15 cm	1.6

From Table 2.3, you can find your water requirements. The above table refers only to the crops grown under irrigation. Crops grown purely under rainfall are not referred here.

2.1.3.1 Example

Suppose you want to irrigate 30 Hectors (1 hector = 10,000 sq m) of a paddy crop. Every round of irrigation for this area has to be covered in 10 to 12 days. Therefore about 3 hectors should be irrigated every day and 30 hectors will be covered in 10 days. Paddy requires 23 cm of depth of water. So the water requirement will be (30 × 10,000 × 0.23 = 69,000 cu.m). If the pump has to work for 10 days, then the pump must supply (69,000/10) = 6,900 cu.m/day. If the pump works for 12 hours a day, then the pump, to be purchased, should supply (6,900/12) = 575 cu.m/hour = (575 × 1000/3600) = 159.7 litres /second. Allowing for seepage, the flow rate can be taken as 175 litres/second. Hence the pump to be purchased must supply a discharge or flow rate of 175 litres/second.

2.1.3.2 Another Example

Suppose you have to irrigate 2 hectors of paddy, 2 hectors of sugarcane, 1 hector of banana and 1 hector of vegetables. What is the capacity of the pump you require? Please proceed to answer questions as shown below (Table 2.4):

Table 2.4

S. No.	Conditions	Paddy	Sugarcane	Banana	Other vegetable
1.	How many acres do you want to irrigate?	2 hectors	2 hectors	1 hector	1 hector
2.	What interval of time is required between two successive irrigations? (read from the Column 2 of Table 2.3)	10 days	10 days	10 days	11 days
3.	How many acres should you irrigate per day on an average? (Divide item 1 by item 2)	0.20 hectors	0.20 hectors	0.10 hectors	0.10 hectors
4.	What will be the depth of water in cms does the crop require? (Read from Column 3 of Table. 2.3)	23 cms	20 cms	18 cms	16 cms
5.	How many cu.m. of water do you require per day? (Multiply Col.3 × Col.4 × 10,000/100)	460 cu.m.	400 cu.m.	180 cu.m.	160 cu.m.

6.	What is the total quantity of water you require per day (add all figures in item 5)	1200 cu.m./ day
7.	You will have now to decide how many hours per day, your pump will work. On that basis calculate the quantity/ minute you require. If, for instance, your pump can work for 8 hours per day, you require	1200/8 = 150 cu.m./hour = 41.7 litres/second

So you have to choose a pump, which will give you 42 litres/second. In this manner you can calculate the quantity you require for your conditions, and on that basis decide the output, your pump should be capable of giving. Any pump manufacturer can recommend a suitable pump taking into account the height to which water has to be pumped and the quantity required.

Some of our esteemed customers frequently ask us, for supply of a particular size of pump and a particular horse power of electric motor either for domestic use or for agricultural purposes. This is not correct. It would be always the best, if our customers estimate at least in a rough way, their probable water requirements before they decide the size of pump set, they require.

2.2 HOW TO CALCULATE TOTAL HEAD OF A PUMP?

In the previous section, we saw how you can calculate the quantity of water you require to be pumped out from your well for meeting your requirements either for domestic purpose or for irrigating your crops. One other factor remains which you should consider, when choosing a pump you require.

Several of our esteemed customers come to us and ask us to supply a pump to be coupled with say a 3 HP motor or 5 HP motor or some other motor. Several sizes of pumps can be driven by a 3 HP motor (e.g. pumps capable of lifting water to a total head of 25 m, 15 m, 11 m and so on). But you will have to take into account your own conditions and calculate what total head the pump, you should have for your conditions. Let us now see what this, term “Total Head” means.

If two persons are asked to throw a ball vertically upward, one person, being strong and possesses more energy, the ball thrown by him upward will go to a greater height than if the other person, who is weak. Thus the height to which the ball will go depends entirely on the amount of upward energy applied to it and the time when the ball leaves the person's hand. The ball goes up only because; a force has been applied to it. At the time, it leaves person's hand, it possesses a certain amount of energy and this makes it possible to travel upward. As the ball moves up, it works against the gravitational force of the earth. In this process, the energy, it initially had, which made it possible for the ball to move up, gradually slows down. When all the energy, the ball possesses, is used, it stops going up. The ball moves up, because the person, who threw the ball upward, had imparted to it some energy.

Now turn your attention to a centrifugal pump. When a pump works, vacuum is created in the suction pipe of the pump. The atmospheric pressure works on the water level in the well and water is pushed inside the suction pipe to fill up the vacuum. In a previous section, we have explained how, it is possible to supply to a height of 5 to 6 m only into the suction pipe under the conditions well explained. We have therefore suggested that as far as possible, the vertical height between the water level in the well and the centre line of the pump should be not more than 5 m.

When the centrifugal pump works, water initially supplied for priming is driven out. As a result vacuum is created in the suction pipe. The impeller of the pump drives the water to the periphery of the pump with much force and water is forced into the delivery pipe of the pump also. So, on the one hand water from the well moves upto the pump through the suction pipe and then water is again forced out by centrifugal force into the delivery pipe. This water gets the energy to move up from the rotation of the impeller of the pump. The height to which the water will move up into the delivery pipe, will depend entirely on the amount of energy, that the impeller of the pump is capable of imparting to the water, as similar to the height to which a ball can travel upward will depend entirely on the amount of energy which the person, who throws it, has applied to it. The amount of energy per kg of water, which the water has, at the time of leaving the pump impeller is usually expressed in terms of the height, to which the water is capable of moving up. This height is called "Total Head" and is measured in metres (m).

Please note that the "Total Head" is an index of the pressure developed by the pump. In actual working, this "Total Head" is utilized for our purposes, namely:

1. for drawing water from the well to the pump through the suction pipe.

2. for driving the water into the delivery pipe.
3. for overcoming friction caused, when water moves with appreciable force through a pipe.
4. for the water struggling its way along sharp bends in the pipe lines, through narrow valves, etc.

You use a pump to lift water from a lower level to a higher level. Let us now imagine that you want to take water from a well to a point, whose vertical height from the well is 15 m. To do this, the pump should be capable of developing a head of at least 15 m of water (ignoring losses). This height is what engineer's call as "Static Head or static height". "Static Head" merely refers to the vertical height to which water has to move up.

But this "Static Head" actually comprises of two parts. The first portion is the vertical height from the water level in the well to the centre line of the pump. This portion is called "Suction lift". The second portion consists of the vertical height from the centre line of the pump to the point at which water is delivered. This portion is called "Delivery Head". You will therefore see that Static Head = Suction lift + Delivery Head. In addition it should be able to overcome resistance arising out of friction, when water flows through some length of pipe lines and through bends, valves etc. Therefore the "Total Head" of a pump means, its ability to supply the rated quantity of water under a particular installation conditions. It should not only lift water up to the static head, but also must overcome all losses.

2.2.1 Calculation of Friction Losses

In measuring the extra head, required to overcome friction etc., we calculate the loss of head. For example, when a heavy box is pushed on a smooth surface, it can move easily. But if the same box is pushed on an uneven (rough) surface, it can be moved with more difficulty because of greater friction. Similarly when water, discharged by centrifugal pump, flows into a pipe, the water experiences friction and this affects its flowing capacity. In other words, this friction results in loss of head, i.e., loss of energy, which the water initially received, from the rotation of the pump impeller.

Now let us consider to what extent there will be loss of head, as a result of water flowing through a pipe line. If you examine the surface of any old pipe under a microscope, you will see how uneven the surface is. There will be many elevations and depressions. When water flows on such a surface it experiences friction. As a consequence, some portion of the energy, that water had, *i.e.*, the "head" is lost. If this loss could be minimized, the pump could raise water to a higher head.

Frictional losses are governed by certain laws. They are: the longer the pipe line, through which water is lifted, the higher is the amount of frictional loss. If the diameter of the pipe line is larger, the frictional loss is less. This is like experiencing lesser traffic jam when vehicles pass through a wider highway. If larger quantity of water flows through the pipe line, the greater will be the frictional loss. For your ready reference, we furnish in Tables 2.3 and 2.4, the loss of head in feet as well as in metres, for different pipe diameters and for the number of gallons/minute (gpm) as well as for the litres/sec (lit/sec) of water that flows through the pipe i.e., the output. You will see from tables that if your pump is capable of giving a discharge of 10 gpm (0.76 lit/sec) and if the diameter of the delivery pipe is one inch, there will be loss of head of 16 feet (4.88 m) due to friction, when the total length of pipe is 100 feet (30.48 m). But if the same quantity of 10 gpm (0.76 lit/sec) flows through a pipe of 2 inches diameter and 100 feet (30.48 m) long the loss is only 0.4 feet (0.12 m). Similarly if 50 gpm (3.79 lit/sec) of water flow through a pipe of 1½ inches diameter and 100 feet (30.48 m) in length, the loss of head will be 40 feet (18.19 m), but if the diameter of the above pipe is increased to 2 inches the loss of head is only 9.4 feet (2.87 m).

Here is how to calculate loss of head due to friction. First find the diameter and the discharge of water you require. Then find from Tables 2.6 or 2.7, what will be the loss of head for the pipe diameter and rated output. This will be for a pipe of (30.4 m) 100 feet long. Calculate the length of your pipe laid in the field. Then work out the frictional loss for your pipe line length.

2.2.2 Loss of Head in Fittings

There is one more source of loss of head, that is, due to the fittings such as bends, foot valve, 'T' piece or sharp elbow, check valve, foot valve and strainer etc. All these are obstructions to the free flow of water, and therefore they entail loss of head. For calculating the loss of head of each kind of fitting, we convert the loss in terms of equivalent pipe length. In Table 2.5, the pipe length equivalent for each kind of fitting and for the various diameters of each of the fittings is given.

Let us take that, you have an already installed pipe of 50 mm (2 inches) diameter with 50 metres of pipe length. Let us assume that, there are 2 bends, one foot valve with strainer, 2 elbows, one non return-valve and a gate valve. The total pipe equivalent for these fittings will be as shown as follows:

For (50 mm)″ diameter	Equivalent pipe length
2 bends at 1.5 m each	3.00 m
1 foot valve and strainer at 3.06 m	3.60 m
2 elbows at 1.62 m each	3.24 m
1 check valve at 0.67 m	0.67 m
1 gate valve 0.67 m	0.67 m
Total	11.18 m

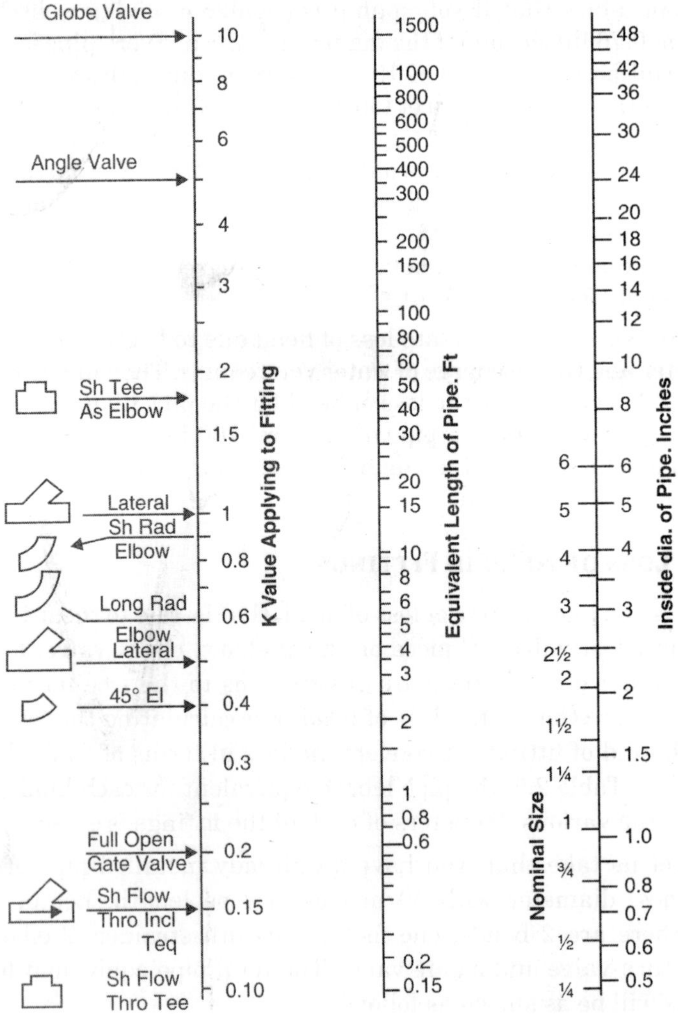

Fig. 2.1: Friction losses in fittings expressed as equivalent length of straight pipe.

Table 2.5: Friction losses of head in feet for each 100′ length of pipe

Discharge in Imp. Gal/ min	Pipe size in inches												
	¾	1	1¼	1½	2	2½	3	3½	4	5	6	7	8
10	50	16	4.3	1.7	0.4	0.13							
20		40	10.0	6.5	1.6	0.5							
30		75	20.0	14.5	3.5	1.1	0.47						
40			37.0	26.0	6.0	2.0	0.8						
50				40.0	9.4	3.0	1.25	0.55					
60				57.0	13.5	4.3	1.80	0.77	0.4				
70				78.0	18.0	5.8	2.4	1.05	0.55	0.17			
80					24.0	7.5	3.1	1.4	0.72	0.22			
90					30.0	9.4	3.9	1.7	0.9	0.28			
100					37.0	12.0	4.9	2.15	1.12	0.35	0.16		
110					43.0	14.0	5.7	2.55	1.3	0.38	0.19		
120					51.0	16.5	6.8	3.0	1.58	0.5	0.22		
125					56	17.8	7.4	3.2	1.7	0.53	0.24		
140					69	22.0	9.1	4.1	2.12	0.67	0.3		
150					80	26.0	10.5	4.7	2.45	0.76	0.34	0.15	
165						29.0	12.0	5.3	2.75	0.87	0.39	0.17	
175						34.0	14.0	6.2	3.25	1.00	0.47	0.2	
180						36.0	14.9	6.6	3.45	1.09	0.49	0.21	

200						45.0	18.3	8.2	4.25	1.33	0.6	0.26	
220						54.0	22.0	10.0	5.20	1.6	0.71	0.31	
225						57.0	27.0	10.1	5.3	1.7	0.75	0.32	
250						70.0	27.5	12.5	6.45	2.05	0.92	0.4	
275							34.0	15.0	7.9	2.5	1.1	0.48	
300							40.0	18.0	9.3	3.0	1.3	0.57	
325							46.0	22.5	10.1	3.5	1.5	0.66	
350							53.0	24.2	12.5	4.0	1.75	0.75	
400									16.3	5.2	2.27	0.99	0.5
450									20.0	6.5	2.8	1.2	0.62
475									21.4	7.4	3.1	1.33	0.68
500									25.0	8.0	3.4	1.5	0.76
600									34.0	11.3	4.8	2.1	1.07
625										12.4	5.2	2.13	1.15
700										15.4	6.5	2.8	1.44

Table 2.6: Friction loss of head in meters for each 30.5 m length pipe

Discharge in lit/sec	Pipe sizes in inches												
	¾	1	1¼	1½	2	2½	3	3½	4	5	6	7	8
0.76	15.24	4.88	1.31	0.52	0.12	0.04							
1.52		12.2	3.05	2.0	0.49	0.15							
2.27		22.4	6.1	4.62	1.07	0.34	0.14						
3.03			11.28	9.93	1.63	0.61	0.24						
3.79				18.19	2.87	0.92	0.38	0.17					
4.55				17.35	4.12	1.31	0.55	0.24	0.12				
5.30				23.77	5.49	1.77	0.73	0.32	0.17	0.05			
6.06					7.3	2.29	0.95	0.43	0.22	0.07			
6.82					9.14	2.87	1.19	0.52	0.27	0.09			
7.58					11.28	3.66	1.49	0.68	0.34	0.11	0.05		
8.34					13.11	4.27	1.74	0.77	0.4	0.12	0.06		
9.09					15.5	5.03	2.07	0.91	0.48	0.15	0.07		
9.47					17.07	5.43	2.26	0.98	0.52	0.16	0.075		
10.61					21.03	6.71	2.77	1.25	0.65	0.20	0.09		
11.37					24.38	7.90	3.2	1.43	0.75	0.23	0.10	0.05	
12.5						8.84	3.66	1.62	0.84	0.26	0.12	0.05	
13.26						10.36	4.27	1.89	0.99	0.31	0.14	0.06	
13.64						10.97	4.54	2.01	1.05	0.33	0.15	0.064	
15.15						13.72	5.52	2.5	1.3	0.41	0.18	0.08	

16.67						16.46	6.71	3.1	1.59	0.49	0.22	0.10	
17.05						17.37	8.23	3.2	1.62	0.52	0.23	0.10	
18.94						21.34	8.38	3.81	1.97	0.67	0.28	0.12	
20.84							10.26	4.57	2.41	0.76	0.34	0.15	
22.73							12.19	5.69	2.84	0.91	0.4	0.17	
24.63							14.02	6.56	3.08	1.07	0.46	0.2	
26.52							16.16	7.38	3.81	1.22	0.57	0.23	
30.31									4.97	1.59	0.69	0.3	0.15
34.10									6.1	1.98	0.85	0.37	0.19
36.00									6.52	2.26	0.95	0.41	0.21
37.88									7.62	2.44	1.04	0.46	0.23
45.46									10.32	3.44	1.46	0.64	0.33
47.36										3.76	1.59	0.65	0.35
53.04										4.69	1.98	0.85	0.44

Thus the above fittings are equivalent to a pipe length of 11.18 m. So, the sum total length will be 50 + 11.18 = 61.18 metres. From Tables 2.7 and 2.8 you can calculate loss of head for 61.18 or say 62 m of pipe length of 50 mm (2 inches) diameter, if you already know the required quantity of water.

Table 2.7: Equivalent length in metres for various pipe fittings

Size in Inches	Bend	Foot value	Tea-joint elbow	Check valve gate valve	Strainer
¾	0.46	0.46	0.41	0.31	
1	0.61	0.67	0.82	0.31	1.22
1¼	0.76	0.76	1.07	0.36	1.22
1½	1.13	0.91	1.53	0.49	1.37
2	1.53	1.22	2.13	0.67	1.53
2½	1.95	1.53	2.74	0.85	1.83
3	2.44	1.83	3.1	1.07	1.83
3½	2.82	2.13	3.66	1.25	2.13
4	3.26	2.44	3.96	1.43	2.44
5	4.27	3.0	5.18	1.86	3.1
6	5.27	3.66	6.4	2.32	3.35
7	6.4	4.27	7.62	2.8	3.66
8	7.62	4.88	9.14	3.35	4.27

2.2.3 Total Head Calculations

The total head of a centrifugal pump is the increase of energy imparted to the liquid flowing from the inlet to the outlet of the pump. This may be expressed in a manometer, which is equivalent to a vertical column of liquid. Accordingly, a centrifugal pump running at a given speed will deliver any liquid, irrespective of its density, to the same height. On the other hand, since the weight of a column of liquid varies with its density, the pressure indicated by the pressure gauge on the pump (expressed in pressure units, e.g. MPa, bar, etc.), will depend on the density of the liquid.

The total head to be overcome by the pump (equilibrium of generated head and system head), is made up of;

1. Total static head
2. The difference between the pressures in suction and delivery tanks

Table 2.8: Minor frictional losses due to fittings expressed as equivalent length of pipe in metres

Diameter (mm)	Ball valve fully open	Valve fully open	Foot valve fully open	Non-return valve	Sieve	90° Elbow	45° Elbow	Tee piece	A/B = 0.25	A/B = 0.5	A/B = 0.75	A/B = 0.25	A/B = 0.5	A/B = 0.75	Entrance into pipe	Exit from pipe
15	4.88	0.11	1.22	1.16	3.05	0.55	0.24	1.04	0.23	0.18	0.12	0.49	0.34	0.13	0.46	0.27
20	6.40	0.14	1.52	1.59	3.66	0.69	0.31	1.37	0.31	0.24	0.14	0.64	0.43	0.14	0.61	0.37
25	8.23	0.18	1.83	1.58	4.27	0.84	0.41	1.77	0.40	0.29	0.18	0.82	0.52	0.18	0.76	0.46
32	11,28	0.24	2.13	2.74	4.88	1.14	0.52	1.79	0.49	0.40	0.24	1.10	0.70	0.24	1.07	0.61
40	13.72	0.29	2.44	3.35	5.49	1.36	0.61	2.74	0.61	0.46	0.29	1.34	0.85	0.29	1.22	0.73
50	16.76	0.38	2.74	4.27	6.10	1.62	0.76	3.66	0.76	0.55	0.37	1.65	1.07	0.37	1.52	0.91
65	19.81	0.43	3.05	5.18	6.71	1.98	0.91	4.27	0.91	0.70	0.43	1.98	1.25	0.43	1.83	1.10
80	25.91	0.53	3.66	5.79	7.62	2.50	1.16	4.88	1.13	0.85	0.49	2.44	1.55	0.49	2.29	1.37
90	28.96	0.61		6.71		2.99	1.37	5.49	1.37	1.00	0.61	2.90	1.83	0.61	2.74	1.59
100	36.58	0.73		7.67		3.66	1.52	6.71	1.52	1.13	0.70	3.35	2.13	0.70	3.05	1.83
125	44.20	0.88		10.06		4.27	1.98	8.23	1.83	1.43	0.85	4.27	2.74	0.85	3.96	2.23
150	51.82	1.07		12.19		4.88	2.29	10.06	2.29	1.74	1.07	4.88	3.35	1.07	4.57	2.74

3. The sum of all the head losses in the system (friction, fittings losses).

The total static head is the difference between the level of the free surface of the liquid on the delivery and on suction sides of the pump, provided that the delivery pipe outlet is submerged. If the pipe discharges above the surface of the liquid, the outlet end of the pipe is to be taken in calculating the difference in level.

The total static head is made up of: (*a*) static delivery head (H_d), which is the vertical distances between the centre line of the pump shaft or in the case of a pump with vertical shaft, the centre of the eye of the lowest impeller and the surface of the liquid in the delivery tank, or the centre of the outlet on the discharge pipe when it is above the surface of the liquid. PLUS, (*b*) the static suction lift (H_{SL}), which is the vertical distance between the level of the liquid in the suction tank and the centre line of the pump shaft or the static suction head (H_{SH}), which is the vertical distance between the level of the liquid in the elevated suction reservoir and the centre line of the pump shaft or in the case of a pump with a vertical shaft, the centre of the eye of the lowest impeller. The H_{SH} is in that case to be considered negative.

$$H_{stat} = H_d + H_{SL} \text{ (or)}$$

$$= H_d - H_{SLH} \qquad \text{...(2.1)}$$

1. The difference in absolute pressures in the suction (p_o) and delivery tanks (p_s), if these are closed vessels.
2. The friction loss is the sum of all losses in head ($H_f = H_{fs} + H_{fd}$) in pipes, bends, valves, instruments for measuring the flow, filters, condensers, heat exchangers, etc., together with the loss of velocity head at exit, corresponding to the exit velocity (v). This last loss becomes relatively more important, the lower the delivery head with large quantity of flow e.g. pumps for land, drainage or other low lift applications. p_3 and p_0 are the delivery and suction reservoir pressures respectively. γ is the specific weight of the pumping liquid

$$H_{tot} = H_{stst} + H_{fs} + H_{fd} + \frac{p_3 - p_0}{\gamma} + \frac{v^2}{2g} \qquad \text{...(2.2)}$$

Example: The following Fig. 2.2 shows a sample pumping installation:

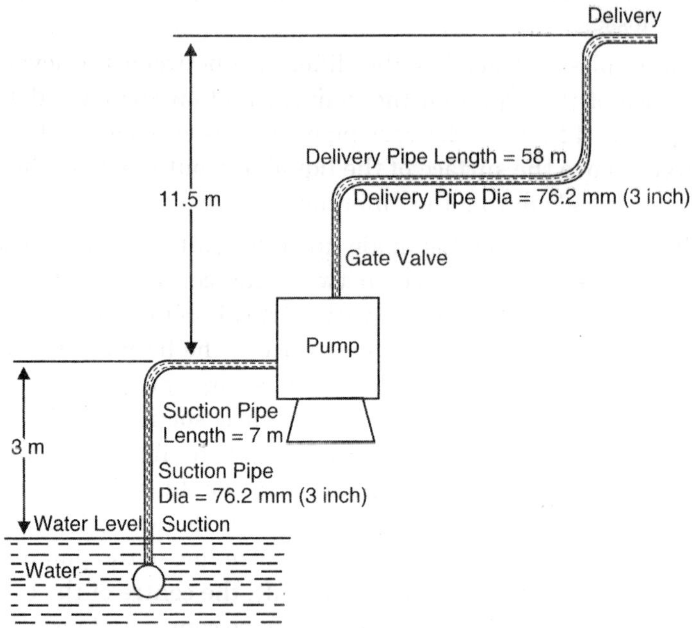

Fig. 2.2: A typical pump installation.

From the Fig. 2.2, you will see that the suction lift is 3 m and the delivery head is 11.5 m. So, total static Head becomes 14.5 m. That means, the pump should raise water to this height of 14.5 m. Water flows through a 76.2 mm (3″) diameter suction pipe of length 7 m and the delivery pipe length of 58 m. So, the total length of the pipe is 65 m. Further the system has one gate valve, 4 bends and one foot valve with strainer. The discharge required is 6.4 lit/sec.

From Tables 2.5 and 2.6, and also from Fig. 2.2, you can read off the pipe length equivalent for the fittings. This is given below:

For 76.2 mm (3") diameter	Pipe length equivalent
For 1 gate valve	0.67 m
For 4 bends at 2,44 m each	9.76 m
For 1 foot valve with strainer	3.60 m
Total equivalent length	14.03 m

The pipe line is already 65 m long. Now add 14.03 m, being the pipe length equivalent for the various fittings. This makes 79.03 m. The loss of head due to friction is deemed to take place in a total pipe length of 79.03 m of 76.2 mm (3″) diameter. For a total quantity of 6.4 lit/sec, from Table 2.4, you will find that the loss of head for a 30.48 m long pipe of 76.2 mm (3″) diameter is 1.06 m. For a length of 79.03 m, the loss of head works out to 2.3 m. So the "Total Head" is equal to the "Static Head" + loss of head due to friction in pipe line and other fittings. In this case Static Head is 14.5 m and the loss of head is 2.3 m making a total of 16.8 m or say 17 m.

From what is stated above, you will see that after deciding what quantity of water you require, you will have to calculate the total head also. For this purpose, you should calculate the Static Head, which comprises of the suction lift and the delivery head. You should roughly visualize how your pump will be installed, what total pipe length will be required and of what diameter. Then you should visualize the number of bends, elbows, 'T' joints, the foot valve, gate valve and/or check valve, etc. You should work out loss of head for all these and finally determine the total head which is static head plus loss of head. We will give you hints, on the selection of the proper pipe size for giving the quantity you want.

It frequently happens that some of you may like to add additional length of pipe, with the already existing pipe line, in order to take water, to some appreciable distance away from the pump. You install a pump and fit pipe lines to the required length. Then you find that the pump does not discharge any water at all, or it may give you very small quantities of water, which is not the quantity you require.

Let us take that the pump available is capable of giving an output of 43 lit/sec, when the total head is 22 metres. (These values are taken from the pump performance graph, supplied by the manufacturer at the time of purchasing the pump). It is found from the performance graph of the pump, that flow rate decreases to 25.8 lit/sec, when the total head increases to 32 metres.

Here are the details of the pipe length to be connected with its existing installation.

Suction side	
Vertical height from water level to centre of pump called suction lift	2.5 metres
Length of 3" pipe on the suction side	3.5 metres
Foot valve and strainer on the 76.2 mm (3") suction pipe line	one
Bend 76.2 mm (3")	one

Delivery side	
Vertical height from centre of pump to point of delivery, called delivery head	11.5 metres
Length of 76.2 mm (3") pipe on the delivery side	58 metres
Bend 76.2 mm (3")	3 Nos.
Gate valve 76.2 mm (3")	one

Earlier in this chapter, we have already explained how you can calculate total head of a pump. We shall calculate the total head for the above installation on the basis of what was explained there.

2.3 PIPE LENGTH EQUIVALENTS

Suction side	
One foot valve and strainer of (76.2 mm) 3" diameter	3.66 metres
One bend (76.2 mm) 3" diameter	2.44 metres
Total for the (76.2 mm) 3" diameter	6.10 metres

So, the equivalent length of suction pipe is 6.1 + 3.5 = 9.6 metres

Delivery side	
3 bends 76.2 mm (3") diameter (2.44 m × 3)	7.32 metres
One gate valve 76.2 mm (3") diameter (1.07 m)	1.07 metres
Total 76.2 mm (3") diameter	8.39 metres

So the equivalent length of delivery pipe is 58 + 8.39 = 66.39 metres.

Since the delivery tank and suction well are open to atmosphere $p_3 = 0$ and $p_0 = 0$. Velocity head at delivery discharge ($v^2/2g$) is neglected at present, since we do not know the exact quantity, the pump is going to give for the total head, which we are yet to calculate.

On this basis we can calculate total head as below:

Suction lift	2.5 metres
Loss of head on the suction side for a total of 9.6 m metres of pipe length of 3" diameter for a discharge of 25.8 litres per second (at 16 metres of loss per 30.5 metres of pipe length)	5.04 metres

Delivery head	11.5 metres
Loss of head on the delivery side for a total of 66.39 metres of pipe length of 3" diameter for a discharge of 25.8 litres/sec (at 16 metres loss per 30.5 metres of pipe length)	34.83 metres
Total head	53.87 metres

This total head of 53.87 metres is far in excess of the maximum total head of 35 metres, for the available pump. Theoretically speaking the pump will not give any output at all, but practically very small quantity will be delivered. This is the reason, why we neglected velocity head, earlier. We suggest, that the diameter of the pipe may be increased on the delivery and suction side, if not at least on the delivery side. The reasons behind this suggestion are explained later.

You will see, from the above calculation that the total head under any given set of conditions of installation is governed by two factors namely (*a*) The vertical height from the water level to the pump and from the pump to the point of discharge and (*b*) loss of head arising from the friction, created, when water flows through the pipe lines, bends, etc. You cannot alter the first factor. But you can alter the second factor by adjusting length and diameter of pipe, number of bends, number of valves etc., in order to ensure that the total head is brought within the operating region of the pump.

You will see, that the frictional losses are more, when the pipe diameter is less. You may then ask why pipes of large diameter cannot be used in all cases. You should remember that pipes of larger diameter cost very much higher than pipes of smaller diameter. A comparative analysis of the present problem will provide the information what we want. (Cost figures are only approximate and actual cost may depend on several factors.)

Table 2.9

Details	For 3" pipe	For 3 ½" pipe	For 4" pipe
Suction side			
Equivalent lengths for			
(*a*) Foot valve and strainer	3.66	4.26	4.88
(*b*) One bend	2.44	2.82	3.26
(*c*) One reducer	–	0.61	1.00
(*d*) Length of the pipe line	3.50	3.50	3.50

Total equivalent length in suction side	9.60	11.19	12.64
Delivery side			
Equivalent lengths for			
(*a*) One gate valve	1.07	1.25	1.43
(*b*) Three bends	2.44 × 3 = 7.32	2.82 × 3 = 8.46	3.26 × 3 = 9.78
(c) One increaser	–	0.61	1.83
(*d*) Length of the delivery pipe line	58.00	58.00	58.00
Total equivalent length in delivery side	66.39	68.32	71.04
Total equivalent length for the installation	9.60 + 66.39 = 75.99 ≈ 76 metres	11.19 + 68.32 = 79.51 ≈ 80 metres	12.64 + 71.04 = 83.68 ≈ 84 metres
For the quantity of 25.8 lit/sec, the loss of head is	63.2 metres	16.9 metres	7.85 metres
Total static head	14.0 metres	14.0 metres	14.00 metres
Total dynamic head (Static head + Friction loss of head)	77.2 metres	30.9 metres	21.85 metres

From the performance graph of the pump, we find that the pump will not give any water, when 3″ pipe is installed. The pump will give 25.8 lit/sec if 3½″ pipe is installed instead of 3″ pipe and will give 43 lit/sec if 4″ pipe is installed instead of 3″ pipe. Correspondingly the power consumption will also reduce, or in other words, the running cost will reduce. Your monthly current bill will be reduced.

Approximate cost in rupees per metre length of pipe is given below (Table 2.10):

Table 2.10

Diameter in inches	Rupees/meter length
½	₹ 62
¾	₹ 74
1	₹ 111

1 ¼	₹ 142
1 ½	₹ 169
2	₹ 217
2 ½	₹ 277
3	₹ 373
3 ½	₹ 454
4	₹ 534
5	₹ 613
6	₹ 766

When we say that a pump has to be fitted with a delivery pipe of 76.2 mm (3″), it does not mean that a pipe of this diameter alone should be used. It only means that it is the best to use a pipe of 76.2 (3″) under normal conditions. You can certainly use pipes of larger diameter, but the cost will be higher. If you use a pipe of smaller diameter you will only create more friction and the discharge of water will definitely decrease. You will see that as a consequence of increasing the diameter of the delivery pipe from (76.2 mm) 3″ to (89 mm) 3½″, the total head has fallen from 77.2 metres to 30.9 metres and for 4″ diameter pipe the total head is reduced from 77.2 m to 21.85 metres. A new pump to suit 3½″ pipe diameter or 4″ pipe diameter condition can be purchased and installed. But our friend has already bought the pump. We cannot go for the purchase of a new suitable pump.

So we are left with the question how best the available pump can be used for our friend's conditions. All this only leads to the thinking, that before you buy a pump, you should consult a competent manufacturer or his agent. You should explain to him in detail, the conditions under which your pump will operate. He will advise you what pump you should buy. This is better than buying some pump and examining how it can be made to work under a given set of conditions. In order to help you to readily decide what diameter of the delivery pipe may be used in Table 2.11.

Normal flow velocity of water in pipes vary from 7 m/sec to 10 m/sec. Under this condition, the friction losses are within allowable limits. If the flow velocity exceeds 10 m/sec, the friction losses increase too much. That's why we do not recommend more than 10 m/sec of pipe velocity. Also we do not recommend less than 7 m/sec, to avoid

Table 2.11: Flow rate through pipe in lit/sec

Pipe diameter		Pipe area	Flow velocity through pipe m/sec							
Inch	mm	Sq. m	3	4	5	6	7	8	9	10
½	12	1.131^{-4}	0.339	0.452	0.566	0.679	0.792	0.905	1.018	1.13
¾	20	3.14^{-4}	0.943	1.26	1.57	1.89	2.2	2.51	2.83	3.14
1	25	4.91^{-4}	1.47	1.96	2.45	2.95	3.44	3.93	4.42	4.91
1¼	32	8.04^{-4}	2.41	3.22	4.02	4.83	5.63	6.43	7.24	8.04
1½	40	1.25^{-3}	3.77	5.03	6.28	7.54	8.8	10	11	13
2	50	1.97^{-3}	5.89	7.85	9.82	12	14	16	18	20
2½	65	3.32^{-3}	9.96	13.3	17	20	23	27	30	33
3	80	5.026^{-3}	15.1	20.1	25.1	30.2	35.2	40.2	45.2	50
3½	90	6.36^{-3}	19.1	25.4	31.8	38.2	44.6	50.9	57.2	63.6
4	100	7.85^{-3}	23.6	31.4	39.3	47.1	55	62.8	70.7	78.5
5	125	0.0123	36.8	49.1	61.4	73.6	85.9	99.82	110.5	122.7
6	150	0.0177	53	70.6	88.4	106	123.7	141.4	159	176.7
8	200	0.0314	94.3	125.7	157.1	188.5	219.9	251.3	282.7	314.2
10	250	0.0491	147.3	196.4	245.4	294.5	343.6	392.7	441.8	490.9
12	300	0.0707	212.1	282.7	352.4	424.1	494.8	565.5	636.2	706.9
16	400	0.1257	377	502.7	628.3	754	880	1005.3	1131	1256.6
20	500	0.1964	589.1	785.4	981.8	1178.1	1374.5	1570.8	1767.2	1963.5
25	650	0.3318	995.5	1327.3	1659.1	1991	2322.8	2654.7	2986.5	3318.3
30	750	0.4418	1325.4	1767.2	2208.1	2650.7	3092.5	3534.3	3976.1	4417.9
36	900	0.6362	1908.5	2544.7	3180.9	3817	4453.2	5089.4	5725.6	6361.7

mud sedimentation, within the pipe. From the above table, you can decide the diameter of pipe, you have to select, based on the quantity of water, you want. This table has to be selected for normal pipe length of 20 m. If you need to layout pipe length for a longer distance say 50 m, then you go for higher pipe size, by which the friction losses are brought within limit. It is usual, that friction losses do not exceed 30 to 35 % of the total static head. If friction losses exceed the limit we select the next higher diameter. For example if you want a quantity of 50 lit/sec, you select 3 inch pipe for the normal pipe length of 20 m. If the pipe length in your site is larger say 100 m, then you select 3 ½ inch pipe diameter and get the same output 50.9 lit/sec. You are likely to get the above discharge even in a pipe of lesser diameter. However, in order that frictional losses may be kept down to the minimum, we have suggested slightly larger pipe diameters. An example is given in the following pages, how to select the correct size of the pipe.

You will also see that the power of your motor is utilized for two purposes namely (*a*) to drive the water into the delivery pipe and (*b*) to overcome frictional losses. For whatever purpose the horse power is used, you have, in any case, to pay current consumption charges. Is it not therefore advantageous for you to use as large a proportion of the Horse Power for getting water instead of allowing that power to be used for fighting frictional losses. So you should always think in terms of keeping frictional losses to the minimum. You can achieve this end by increasing pipe diameter and avoiding the use of too many bends etc. When another opportunity occurs we will discuss in these pages in more detail what percentage of Horse power is used for each of the above two purposes under a given set of conditions and how a larger share of that power can be utilized directly for lifting water. In the meanwhile the Table 2.11 will give you guidance in a very general way of what pipe diameter can be used for obtaining a given output of water for different lengths of pipe lines.

2.4 WHAT IS HORSE POWER?

You have heard the term "Horse Power" being used frequently. Have you thought what it precisely means? Here is something for you to think.

A "horse power" is a measure of the rate at which work is done. Do you find it difficult to visualize what work is? Well work is what most of us do physically. Suppose you have, before you, a stone weighting 10 kgf. If you want to lift it from the ground level to a height of say 1 metre, you have to do work. Please note that when work is done, a weight of material (in this case a stone weighing 10 kgf) should be moved. Unless this movement actually takes place, it cannot be said that work has been done. In the above example, work amounting to 10 kgf. m is done because a weight of 10 kgf has been moved up by 1 m (10 kgf × 1 m = 10 kgf. m).

Suppose you have a well, and you lift 10 litres of water from the well to a height of 30 metres. One litre of water weighs 1 kgf. So 10 litres of water weigh 10 kgf. The work done in this case is 30 m × 10 kgf or 300 kgf. m.

In both these examples we have considered only two factors namely the weight of the material moved and the distance to which it has been moved. We have not considered the time factor that, what time has been utilized to complete the work. This is also equally important.

Let us take the case of two men—one a very strong man and another relatively weak person. Suppose we ask each of them to lift a box,weighing say 10 kgf. They are asked to lift this box to a height of 1 metre. The strong man may do it quickly say in 3 seconds. The weak man may struggle and do the work in 10 seconds. The work done in both cases is 10 kgf × 1 metre = 10 kgf. m. Since the strong man has done this work in 3 seconds, the work he does in one second is 10/3 or 3.33 kgf. m/sec. The weak man does work at 10/10 or 1 kgf. m/sec. The work divided by time is called power and is measured in kilowatts (Horse Power).1 Watt = 1 Newton. m sec. and 1 kW = 1000 watts = 1000 Nm / sec = 102 kgf.m/sec, since 1 kgf = 9.81 Newton.

In early days, in England, work was done by horses in a larger way. So the power of the horse was taken as the standard. British people thought that a good horse could do work at the rate of 550 foot pounds per second. This they called as Horse Power (HP), which means that in one second 550 foot pounds of work should be done. As and when power through steam or electricity was used for doing work, the same measure of work was applied. Any machine, which can do 550 foot pounds of work in one second is said to have one Horse power. i.e., 1 HP = 550 ftlb/sec. An average human may perhaps do work equal to 1/8 to 1/10 of a Horse Power. 1 HP = 746 watts (W) or 0.746 kilowatt (kW).

Suppose you fit an electric motor to a pump and use the pump to lift water from your well. Let us also assume that the pump has to lift water to a height of 30 feet. One Horse Power means 550 foot pounds per second. Your electric motor of one Horse Power would do 550 foot pounds × 60 seconds or 33,000 foot pounds of work if it works for 60 seconds or one minute. Since the height, to which water lifted is 30 feet and since one gallon weighs 10 lbs, the above work will give you 33,000/30 × 10 or 110 gallons per minute. In actual practice the quantity of water we may get may be very much less than 110 gallons because of frictional losses in the pipe lines and many other factors. This loss may be as much as 40 to 50%.

2.5 HOW TO CALCULATE THE REQUIRED POWER (KW, HP) OF MOTOR?

Now you know the total head of the pump and the required quantity. From these, the power in kW or in HP, required by the motor, you can calculate, by applying the formula:

Power (N_o) (in kW) =

$$\frac{\text{Specific Weight } \gamma \text{ in N/m}^3 \times \text{Quantity Q in m}^3\text{/sec} \times \text{Head H in meters}}{1000}$$

Since 1 m³/sec = 1000 lit/sec. Specific weight of water = 1000 kg/m³ = 1 gm/cc = 9.81 × 1000 N/m³ 1 HP = 0.746 kW

For a quantity of 6.4 lit/sec and a total head of 17 M

$$N_o = \frac{\gamma QH}{1000} = \frac{9.81 \times 1000 \times \dfrac{6.4}{1000} \times 17}{1000} = 1.07 \text{ kW (1.4 HP)}$$

It is always necessary to allow an extra power of about 10 to 15 %. This will make the motor to run safely even when the motor is overloaded due to unavoidable circumstances such as change in suction head due to season, etc. Allowing 10% overloading capacity, you will have to go in for a motor of 1.07 × 1.10 $\doteq$ 1.18 kW (1.58 HP). The nearest motor available is with 1.5 HP. Any pump manufacturer can supply you a suitable pump capable of giving an output of 6.4 lit/sec at a total head of 17 m and with a motor power of 1.5 HP.

2.6 INSTALLATION OF PUMPS: A CUSTOMER'S PROBLEM

An esteemed customer writes that he has a well at the bottom of a hill. The water depth in his well is about 8.5 m, in the winter, and 5.5 m in summer, i.e., water level in the suction well varies by 3 m, due to seasonal variations. The pump is to be installed at 3.5 m above the water level in the well during winter season and the pump is 3.3 m below the ground level. He has kept a collecting water tank about 37 m away from the well, on the top of a hill, which is 600 m high from the mean sea level. But the vertical height from the ground of the well to the bottom of the tank is 4.8 m and the tank is 4 m high. The quantity he wants is 9.47 lit/sec. He gives below a rough diagram of the proposed water tank, and the site condition.

Our customer has purchased a 3.7 kW (5 HP) pump with an electric motor from a distributor. He wants to know whether the pump will work efficiently and give the rated discharge or not.

2.6.1 Suction

Theoretically any centrifugal pump situated at mean sea level can suck water up to 10.336 m. But in actual practice there are frictional losses in the suction pipe, losses in valves, bends etc., and the location of the site, where the pump is to be installed. Also an allowance has to be given for vapour pressure effect. Under normal condition, for agricultural irrigation drainage and other similar purposes, where the pumping water temperature is around 15°C, the allowance for vapour pressure is 0.336 m. Accounting all, the static suction height for a pump can be assumed and taken as 5 to 6 m.

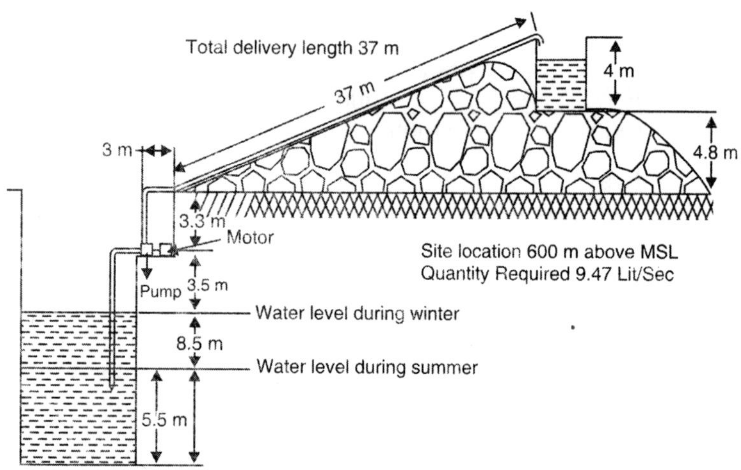

Fig. 2.3

The customer's village is situated at an elevation of 600 m above mean sea level. Hence there will be a corresponding reduction in atmospheric pressure. The suction head of the pump also correspondingly reduces. There will be a reduction of 1 m in atmospheric pressure for every 900 m of height in altitude. Thus; the loss in atmospheric pressure will be 0.67 m for an elevation of 600 m above mean sea level. This will definitely affect the suction lift of the pump.

In any well, the water level will vary from season to season. In his well, the difference between the water levels in summer and in the rainy season is 3 m. You should make allowance for this also. Taking into account (*a*) the suction and delivery heights (*b*) loss of head due to friction (major and minor), (*c*) loss of head, due to the elevation of the place above the mean sea level (reduction in atmospheric pressure), and (*d*) the difference in water level, in the sump, due to seasonal changes, the suction height should be calculated.

Theoretical suction height (= atmospheric pressure) at mean sea level	10.336 m
Reduction of head due to the elevation of the place	0.670 m
So the actual atmospheric pressure or suction lift available	9.666 m
Reduction of head due to vapour pressure	0.336 m
Friction losses (major and minor)	2.270 m
Water level change due to season	3.000 m
Total allowance	5.606 m

So the allowed suction lift in winter will be 9.666 – 2.606 = 7.060 m and the same in summer will be 9.666 – 5.606 = 4.060 m. Please note that the length of pipe extended into water to avoid air lock during running of the pump is also added here. It should be added along with the length of suction pipe for calculating friction losses. Since the delivery tank and suction well are open to atmosphere $p_3 = 0$ and $p_0 = 0$. Velocity head at delivery discharge ($v^2/2g$) is neglected at present.

In other words vertical distance between the pump and the water level at its lowest point should not be more than 4.06 m in winter and 7.06 m in summer. Otherwise, the pump will not lift water. But our customer's well has only 3.5 m depth during winter and 6.5 m during summer. Hence pump will work safely as far as the suction conditions are concerned.

2.6.2 Delivery

The pump is installed below the ground level i.e., inside the well. The height from the pump, to the ground level, is (3.3 m) 10 feet. The vertical height from the ground level to the tank bottom is (4.8 m) 15 feet. The tank height is (4 m) 13 feet. This brings the total height on the delivery side to (12.1 m) 38 feet. The suction height is (6.5 m) 21.5 feet. Adding this height along with delivery height, we get the total static height as (18.6 m) 59.5 feet.

We have to take into account the loss of head in taking the water over a distance of (37 m) 122 feet. For this length and for an output of (9.47 Lps) 125 gpm, the loss of head due to friction will be (21.5 m) 68.3 feet, if a (50 mm) 2 inch diameter pipe is used. If this too is added, we get the total head as (18.6 + 21.5 = 40.1 m) 59.5 + 68.3 = 127.8 feet. This does not take into account loss of head due to friction in valves, bends etc. If we take minor losses also into account, the total head will be still higher. Also please note that during the actual course of installation, these measurements increase slightly and the total head may exceed still further. Since the total available head in the pump, the customer has is (26 m) 85 feet only, there is no margin left. This will result in the pump not giving any discharge, theoretically, but actually the pump will give a very small quantity of water and its working will become uneconomical.

If a delivery pipe of 3 inch diameter is used instead of 2 inch diameter, the loss of head will be reduced to (2.8 m) 9 feet only. The total working head now will be 18.6 + 2.8 = 21.4 m (say 22 m) or 61 + 9 = 70 feet. Now the pump will definitely give the quantity of (9.47 Lps) 125 gpm, since the total head does not exceed the rated head of the pump 26 m (85 feet). Actually the pump will give more discharge than 125 gpm, since the total is less than (26 m) 85 feet.

We therefore recommend the use of a 3″ pipe in this case. If this is done the pump will give an output of not less than (9.47 Lps) 125 gpm.

2.7 A CUSTOMER'S DILEMMA—A PROBLEM IN PUMP INSTALLATION

The other day one of our esteemed customers stepped into our Sales Department and asked: "I purchased a centrifugal pump, 4″ × 3″ coupled to a 7.5 HP Electric Motor. After erection I find that it gives very low output of water. What shall I do? After further consultations with him, the following facts came to light:

The customer found that at the site where the pump set was proposed to be installed, the suction lift is about (4.51 m) 15 feet. He therefore thought that the suction pipe may be (6.1 m) 20 feet long. The length of the delivery pipe at his site is 183 m (600 feet). His neighbour had fitted a 3″ suction pipe and 2″ delivery pipe to his 4″ × 3″ pump. Our customer wanted to do the same because his pump and his neighbour's were identical in type and capacity. As compared with a 2″ pipe, a 3″ pipe was 15 to 20 percent more costly and a 4″ pipe was 100 per cent costly. The customer therefore estimated that in his case, using (6.1 m) 20 feet length of suction pipe of 3″ instead of 4″ and (183 m) 600 feet length of delivery pipe of 2″ instead of 3″ would save him about ₹ 2000 to ₹ 2500 in all. He used pipes of smaller diameters and the result was that the pump was supplying hardly (6.29 lit/sec) 83 GPM. This is a kind of problem which is faced with, very often.

In Fig. 2.4(*a*), a sketch of the installation made by our customer is given and in Fig. 2.4(*b*), a sketch of his neighbour's installation is given. In both cases centrifugal pump 4″ × 3″ has been used.

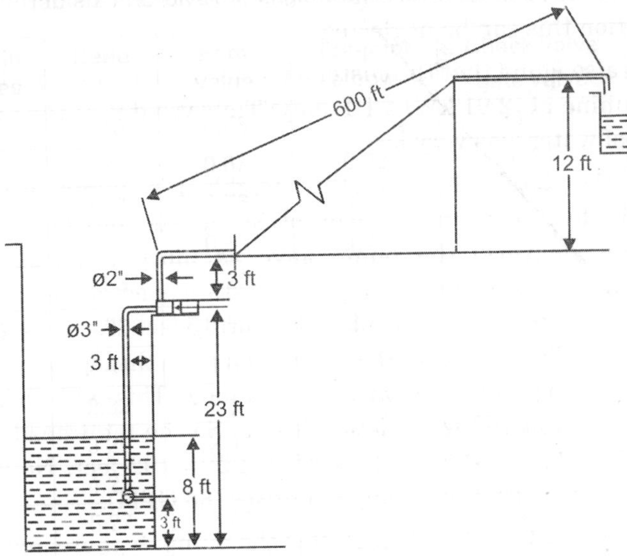

Fig. 2.4(*a*)

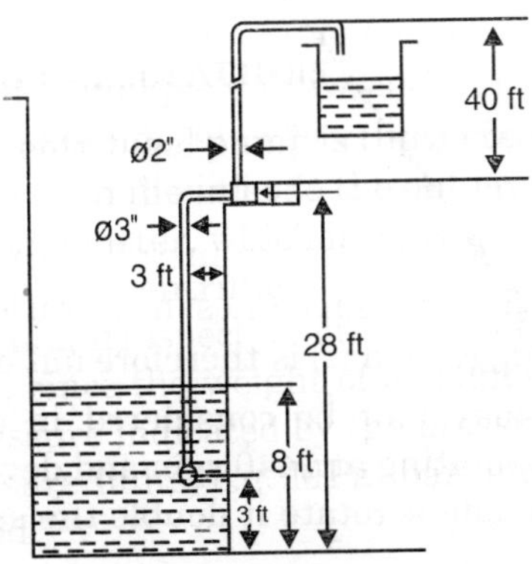

Fig. 2.4(*b*)

According to our customer, his pump was giving a discharge of only (6.3 lit/sec) 83 GPM whereas his neighbour's has got a much more ample discharge though both of them had used 3″ pipe for suction and 2″ pipe for delivery.

The main point here is that in our customer's case the suction lift is (4.55 m) 15 feet and the delivery head is about (4.55 m) 15 feet with the delivery pipe length of (183 m) 600 feet. In his neighbour's case the suction lift is (6.1 m) 20 feet and the total delivery head is (12.2 m) 40 feet. In the neighbour's installation the delivery pipe is not very long and therefore there is no appreciable loss of head on this account. For our calculation this can be neglected.

It was also found that the customer's electric motor of (5.7 kW) 7.5 HP was running at (2.91 kW) 4 HP only. This was due to the fact that the output of water was very low.

Since the total head in the neighbour's pump was only (18.2 m) 60 feet ((6.1 m) 20 feet of static suction, (12.2 m) 40 feet of static delivery) the pump gave as much as (18.2 lit/sec) 240 GPM. In the case of our customer the static suction head is (4.55 m) 15 feet and static delivery head is (4.55 m) 15 feet. To this should be added, the loss of head by allowing the water to run through a relatively narrow 2″ delivery pipe (183 m) 600 feet long. This will amount to as much as (49.4 m)162 feet friction losses. So the total head will be (4.55 + 4.55 + 49.4 = 84.6 m) 15 + 15 + 162 = 192 feet. No wonder therefore that the output was only 6.3 lit/sec (83 GPM).

The customer has already installed the pump. Is there any way of getting more water? The following suggestions are offered (*i*) Change the delivery pipe to 4′ diameter instead of 2″ pipe, in order to reduce the loss of head due to friction or, (*ii*) Abandon the direct coupling and use a belt drive. The reasons will be explained later on in this note. This will involve use of a pulley of a suitable size which would further increase the speed of rotation of the impeller.

Since our customer has already installed the pump, it would be difficult for him to discard the pipe lines; he has used and provide pipes of larger diameter. Suggestion (*i*) is therefore out of question.

Suggestion (*ii*) may now be considered in detail. Suppose the impeller of a pump is rotating at 1440 rpm and develops a total head of 35 feet. Suppose we made it rotate it double the speed, say 2880 rpm. Engineers tell us that the head, it will develop now will be $\frac{2880^2}{1440^2} \times 35$ $= 4 \times 35 = 140$ feet. Thus by doubling the speed at which the impeller rotates we will increase the head to four times. This will require a larger driving force and so a greater horse-power will be required in the motor. Engineers again tell us that if 1 HP is used when running the impeller at 1440 rpm, the horsepower required to run it at 2880 rpm will be $\frac{2880^3}{1440^3} \times 1 = 8 \times 1 = 8$ HP. In the present case, our customer had found that the 7.5 HP motor was running only at 4 HP. The balance of 3.5 HP left can well be used for increasing the speed. This will provide him more head and result in greater output of water. But in this case it should be emphasized that appreciable amount of HP is wasted to fight the frictional losses arising as a consequence of using 600 feet of pipe lines of 2″ diameter. This is a contingency which we should avoid even when a pump is installed and to this end proper technical opinion will have to be obtained in advance.

2.8 INSTALLATION OF PUMPS: SOME COST CONSIDERATIONS

We have seen how loss of head can be minimized if the diameter of pipe lines is increased. It is necessary to consider some aspects of cost. You have to incur expenditure (*a*) on the pump and motor, (*b*) on the pipes and (*c*) on electricity required to run the pump set. You will agree that a larger amount of electric power will be consumed if you use a motor of a higher horsepower. If a motor of a lower horsepower can supply all the water you require, where is the need to go in for a higher horsepower motor for which you will have to pay more on electric current? Further, pipes with a larger diameter are relatively very costly. Perhaps you go in for larger diameter pipes merely to reduce frictional losses. This means that you will have to invest more money on them. If you use larger diameter pipes frictional losses will be less. Then you may go in for a lesser horsepower motor. This will save you expenditure on current. This saving may make up the increased initial expenditure you may incur on buying pipes of larger diameter. You will thus see that if you have to economize expenditure on your pump installation, you have to find out which is the most profitable kind of installation.

We shall consider this matter with reference to a case we recently met with. One of our customers has installed a pump 3″ × 2″ coupled with a motor of (9.4 kW) 12.5 HP. This pump is capable of giving a discharge of (18.94 lit/sec) 250 gallons per minute at (21.34 m) 70 feet total head. This will reduce to (12.12 lit/sec) 160 gallons per minute as the head increases to (35.6 m) 120 feet. His requirement is (12.12 lit/sec) 160 gallons of water per minute. His installation is given in the diagram. It will show that the static lift on the suction side is (3.1 m) 10 feet with a 3″ pipe line. On the delivery side, the vertical height or static head is (3.1 m) 10 feet and the pipe line is (106.7 m) 350 feet long with a diameter of 2″. For the purpose of our calculation we may ignore losses due to valves, bends, etc. Let us now calculate the total head of the installation.

Static lift on suction side	(3.10 m)	10 feet
Loss of head on the suction side for 10 feet pipe of 3" (Loss of head of 12 feet for every 100 feet on pipe length for an output of 160 gallons per minute)	(0.37 m)	1.2 feet
Static head on delivery side	(3.10 m)	10 feet
Frictional loss for 350 feet of pipe line of 2" diameter for an output of 160 gallons per minute (Loss of 29 feet for every 100 feet of pipe length)	(30.90 m)	101.5 feet
Total	(37.47 m)	122.7 feet

Since the total head is more than (36.6 m) 120 feet (disregarding the fittings), the pump is not likely to give (12.12 lit/sec) 160 gallons per minute. You will see from the above calculation that the vertical height of the installation is only (6.1 m) 20 feet. Another (31.3 m) 102.7 feet comprises frictional losses. In other words proportion of (20/122.7) or say 1/6 of the electric power consumed by the motor is being used for lifting water. The balance of 5/6 of the power is used to fight friction. What a waste!

Let us now assume that we substitute a 3" diameter pipe for the delivery side of (107 m) 350 feet. The total head will now work out as follows:

Static lift on the suction side	(3.10 m)	10 feet
Loss of head on the suction side as before	(0.37 m)	1.2 feet
Static head on the delivery side	(3.10 m)	10 feet
Frictional loss for 350 of 3" pipe for an output of 160 gallons per minute (Loss of 12 feet for every 100 feet of pipe length)	(12.80 m)	42 feet
Total	(19.30 m)	63.2 feet

You will see that the total head is now reduced to (19.3 m) 63.2 feet only as against (34.47 m) 122.7 feet previously. A pump 3″ × 2″ is suitable for a minimum head of (24.4 m) 80 feet with a (9.4 kW) 12.5 HP motor and it will give (18.94 lit/sec) 250 gallons per minute. You do not want so much quantity. So why should you use a (9.4 kW) 12.5 HP motor? Since you want only (12.12 lit/sec) 160 gallons per minute, you can use another pump 3″ × 3″ with a (5.5 kW) 7.5 HP motor. This will give (18.94 lit/sec) 250 gallons per minute at (15.24 m) 50 feet head and (9.47 lit/sec) 125 gallons at (26 m) 85 feet. It would give (12.8 lit/sec) 169 gallons at about (19 m) 62 feet. You will thus see that the horsepower has been reduced from (9.4 kW to 5.5 kW) 12.5 to 7.5 by increasing the diameter of the delivery pipe. In this way we can save current charges. Here the frictional loss is only (13.2 m) 43.2 feet out of the total head of (19.3 m) 63.2 feet.

Let us take a third alternative. Suppose we increase the delivery pipe diameter to 4″ keeping the suction pipe diameter at 3″. The resulting total head will then be:

Static lift on the suction side	(3.10 m)	10 feet
Loss of head of suction side as before	(0.37 m)	1.2 feet
Static head on the delivery side	(3.10 m)	10 feet
Frictional loss for 350 feet of 4" pipe for an output of 160 gallons per minute (Loss of 2.75 feet for every 10 feet of pipe length)	(3.00 m)	9.6 feet
Total	(9.57 m)	30.8 feet

If you want (12.12 lit/sec) 160 gallons per minute for a total head of (9.5 m) 31 feet, you can use another pump 3″ × 3″ coupled with a (2.2 kW) 3 HP motor. For this pump you can use of 3″ × 4″ increaser at delivery line and then (106.7 m) 350′ length of 4″ delivery pipe. It will give an output of (12.9 lit/sec) 170 gallons at (10.7 m) 35 feet head. You will see that a further increase in diameter of the delivery pipe has resulted in a further reduction of the required horse power for the same output.

The three installations referred to above are compared in the following statement (Table 2.12):

Table 2.12

Details	First installation	Second installation	Third installation
Diameter of suction pipe	3 inches	3 inches	3 inches
Delivery pipe	2 inches	3 inches	4 inches
Static lift	10 feet (3.1 m)	10 feet (3.1 m)	10 feet (3.1 m)

Static head	10 feet (3.1 m)	10 feet (3.1 m)	10 feet (3.1 m)
Total head (including loss of head due to friction)	122.7 feet (37.4 m)	63.2 feet (19.3 m)	30.8 feet (9.4 m)
Horse power (kilowatt)	12.5 (9.4)	7.5 (5.5)	3 (2.2)
Output of water	160 gallons per minute (12.12 lit/sec)		

Let us now go into the question of costs. Here are the details (Table 2.13):

Table 2.13

1.	Cost of pump and motor	₹ 28.000	17,000	12,000
2.	Starter & Switch	₹ 8,200	7,600	4,000
3.	Cost of 375 feet of pipe length	₹ 24,803	42,634	61,036
Total initial cost		₹ 61,003	67,234	77,036

Number of units consumed per month	5,625	3,375	1,350
Cost of power per month @ ₹ 3.50 / unit	19,687.5	11,812.5	4,725

(To get kilowatt hours multiply HP × 0.746. It is assumed that a pump works for 300 hours per month. Cost of power is taken as ₹ 3.5 per kilowatt hour).

You will see that the third installation involves an additional initial expenditure of ₹ 16,033 only as compared with the first but the current consumption charges are very low. This comes to comparing between the first and third you will naturally choose the third. Even in the first month you have made a good savings in the higher initial investment of ₹ 14,962.5 on current charges. The second installation involves a reduced savings of ₹ 7,875 in one month as compared with the first. You would therefore prefer the second installation over the first and between the second and third installations over the third. As between the second and third installations the initial cost in the third is more by ₹ 9,802, but you save current charges by ₹ 11337.5 per month. You can make up the initial extra expenditure in 2 to 3 months. Beyond this period it is all saving for you.

You will therefore see that the first installation is very costly from every point of view. You are sure to select the third installation because in the long run it would prove economical. You will also note

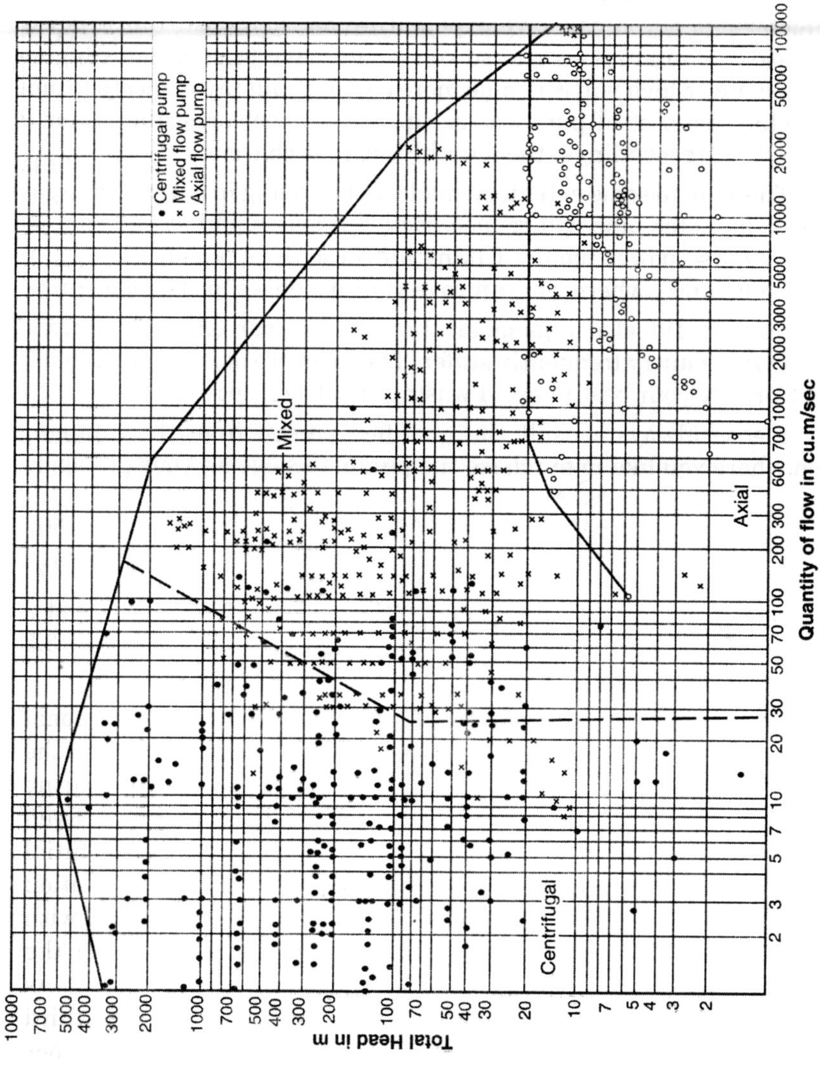

Fig. 2.5: Pump selection chart.

that the large initial expenditure of the installation is due to the heavy expenses on the pipe lines.

All this leads to the inference that you should clearly visualize the installation you want to have. Then you should work out costs and go in for the most economic kind of installation. If you use small diameter pipes, you can save expenditure on their cost. But this will increase frictional losses and so you would have to go in for a higher power motor which will consume more current. But if you increase pipe diameter, the cost of pipe will go up, but the power of the motor is reduced, correspondingly the monthly current charges.

This can, to some extent, be made up by saving in current charges. Larger diameter pipe will reduce frictional losses and so you can use a lower horse power motor. Taking all these, it is possible for you to strike a balance, where you can save your expenses in the long run.

You should also remember that electric current is a precious thing. Our country frequently suffers from power shortage. You should therefore try your best to avoid wastage of electric power.

Please always consult the qualified person from the company, from where you like to purchase the pump, to solve all your problems.

3 INSTALLATION

3.1 INSTALLATION OF PUMP SETS

Are you one of those who has purchased a pump set or about to purchase one? You would naturally like to know how to proceed with its installation. Here are a few tips which will help you to do your job perfectly.

3.2 PRELIMINARY CHECK

1. Check up the machine at the time when you take it over from the Railways or the Lorry Transport Company, who transported the pump set. See that the package is intact. All parts of the pump are not broken, and that there are no deficiencies as compared with the packing list, which the manufacturer might have sent. If there are damages or deficiencies, make claims to the appropriate person. This should be done without delay.
2. If everything is all right, remove the varnish on the shaft. The manufacturer has given the varnish only for protection against rust.
3. Check up all the places, where grease has been applied by the manufacturer. If there is dust or other foreign matter, remove the dusty grease or the foreign matter completely and apply grease afresh. The important point is that every part, which requires grease, should have it without fail and that too without the grease getting mixed with dust or other foreign matter.

3.3 LOCATION

Install the pump set in a dry and clean location which has good lighting. Dirt ultimately causes damage to pump and motors. In order to facilitate dismantling and repairs, there should be adequate lighting and head room space. In locations where flooding is possible, select two locations—one at a lower level for use when flood has receded, and another at a higher level for use when there is flooding. You will have to transfer the pump to one of these locations as circumstances demand.

3.4 FOUNDATION

The Foundation should be heavy to reduce vibrations and should be rigid enough to avoid any twisting or wrong alignment. A space of 25 to 40 mm is allowed between the base plate and the top of the foundation which is filled with grouting to secure a uniform load distribution.

Each foundation bolt should be surrounded by a pipe sleeve, three or four diameters larger than the bolt. After the concrete mixture is poured, the pipe is held solidly in place while the bolt may be moved to conform to the corresponding hole in the large plate. When a unit is mounted on steel work or other structure, it should be placed as near as possible to the main member, beams and walls and be supported so that the base plate will not be disturbed.

3.5 ALIGNMENT

The alignment made by the manufacturers of pumps may get disturbed during transport. A correct alignment is required again at site of erection. First remove the coupling bolts. Place the pump and motor under assembled condition on the base plate over the foundation. Raise the base plate and until there is gap of 20 mm (¾") to 40 mm (1½") between the bottom of the base plate and the top of the foundation. Then adjust the level of the pump shaft. If necessary wedges (or shims as they are called) may be placed under the base plate to adjust level, to make the base to be horizontal. Rotate the pump shaft and driver shaft separately by hand in order to see that they move freely. Now insert one pin of the coupling so that both halves of the coupling can be rotated at the same time. Fix a strong metal bracket under the nut of the coupling pin. The shape and method of attachment of this bracket are shown in Fig. 3.2.

3.5.1 Axial Alignment

Lay a straight edge across the rims or hub tops of the two half couplings and check radial alignment at 90° interval. Both couplings should be rotated and readings checked at 90° interval. Check the radial gap as shown, again at 90° interval and rotating both couplings.

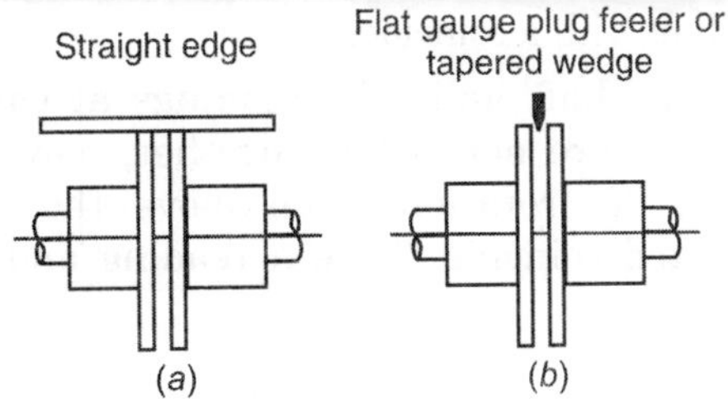

Fig. 3.1: Axial alignment.

Any repositioning of the motor to correct for radial misalignment must be followed by a further check on gap clearance which should be carried out again at four points, at 90 to each other using either a taper or a parallel gauge in conjunction with feelers.

3.5.2 Angular Alignment

1. Clamp two dial gauge to the coupling halves as shown in Fig. 3.2 and zero the gauges.
2. Link the coupling halves or mark them as previously described.
3. Rotate both shafts through 180 and note the gauge readings. If both gauge readings are the same (but not necessarily zero) the angular alignment in the vertical plane is correct.
4. Rotate both shafts through a further 90 and note the gauge readings. If the readings are the same (but not necessarily zero), the angular alignment in the horizontal plane is correct.

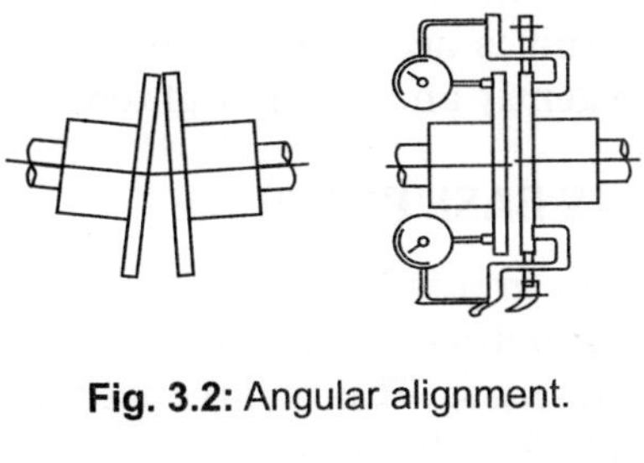

Fig. 3.2: Angular alignment.

3.5.3 Parallel Alignment

1. Clamp a dial gauge to the coupling halves as shown in Fig. 3.3 and zero the gauge.
2. Link the coupling halves so that both shafts can be rotated together. If the shafts cannot be linked together, mark each coupling half with a coincident line.
3. Rotate both shaft and take readings at each 90° rotation. If the shafts are not linked together, move first one shaft through a 90° rotation, then move the other until the lines marked coincide. Take a reading and repeat at each 90° rotation.

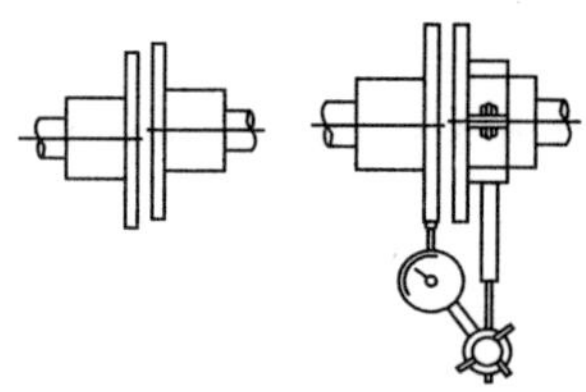

Fig. 3.3: Parallel alignment.

Fix a dial gauge to the bracket in such a way that it touches the outer periphery of the other half of the coupling. If a suitable dial gauge is not available the bracket should be shaped to a pointed nose. This point should be bent down so that it comes within a few thousandths of an inch of the rim of the opposite half of the coupling. If a dial gauge is used, the variation will be read directly from it. If not the variation must be measured by a feeler gauge. Wherever a feeler gauge is not available, papers of various thicknesses may be used. Rotate the shaft by hand. As you rotate note the variation in alignment either in the dial gauge, or feeler gauge or with pieces of paper at various points namely (*a*) at the start, (*b*) after rotating 90 degrees, (*c*) after rotating 180 degrees and (*d*) after rotating 270 degrees. If there is variation, adjust the base plate level or the position of the motor and pump till a perfect alignment is attained. Then all the coupling pins may be tightened. After the unit is accurately leveled and aligned, the hold-down bolts should be gently and evenly tightened before grouting.

3.6 GROUTING THE BASE PLATE

The base plate is then grouted before giving pipe connections. The purpose of base plate grouting is to prevent lateral shifting of the base plate and to increase its mass. This will reduce vibration and also even out irregularities in the foundation.

The usual mixture for grouting a pump base is composed of one part of pure Portland cement and two parts of building sand with sufficient water to cause the mixture to flow freely under the base (a heavy cream consistency).

In order to get the grout properly at all points under the base plate, use a funnel made of tin. Pour the grout through the funnel until the entire space under the base plate is completely filled. A stiff wire could also be worked through the grout holes in the base plate to stir the grout and see that it is free of all air pockets. The exposed surface of the grout should be covered properly with a gunny to prevent cracking from drying too rapidly. After the grout is sufficiently set which takes about 48 hours, the wooden frames may be removed. Then the exposed surfaces of the grout and foundations are finished to a smooth surface. When the grout is hard (72 hours or more), the hold-down foundation bolts should be finally tightened and the coupling halves rechecked for alignment.

3.7 FURTHER CHECK OF ALIGNMENT

Fit all the pipe connections. Before working the pump, again rotate the shaft by hand and test the couplings with a dial gauge, or feeler gauge or by using paper bits. If any adjustments are required carry them out. The unit is now ready for working.

3.8 PIPE WORK INSTALLATION

The suction and discharge pipes and all associated valves and similar equipments should be supported and anchored near the pump, but independent of it so that no strain is transmitted to the pump casing. Such strains may cause damage to the pump or distortions and rubbing of pump parts.

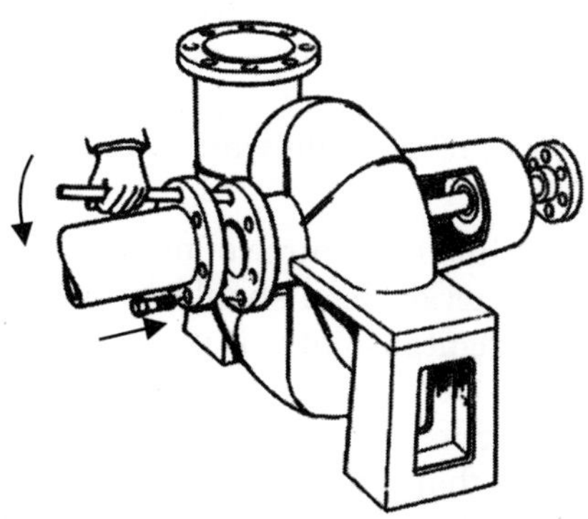

Do not overstrain the pipework when connecting to pump

Fig. 3.4: Suction pipe installation.

It is essential that no load be imposed on the pump parts, as this is liable to disturb shaft and bearing alignment resulting in excessive wear and possible seizure. The pipe work should, therefore be independently supported and anchored against pipe forces. When erecting the pipe work, the pump flanges should be the final closure joints. Check coupling alignment before and after the pump flanges have been tightened. Any alteration in alignment is indicative of pipe work loading on the pump branches and steps must be taken to remove the load.

3.8.1 Suction Pipe Work

Wherever possible, the pump must be made to operate on a positive suction. Eliminate all possibility of air pockets being trapped in the suction pipe work, e.g., ensure pipe work rises continuously towards the pumps; where pipe diameter is reduced in direction of flow use flat topped taper pieces.

The diameter of the suction pipe should be sized according to the flow and allowable head loss and may not, therefore, be the same diameter as the pump suction branch.

A straight length of pipe adjacent to the pumps suction branch, **a length about 4 to 6 diameter of suction pipe** is desirable. Where a bend is unavoidable, use long bends and not elbows.

The suction pipe should be accessible and not embedded in concrete.

When choosing foot valve, strainers, bends, etc. select those which will provide minimal restriction to flow.

Care must be taken in the design of the suction inlet. Ideally the pipe end should be at a depth of at least **five pipe diameter** below the water surface, and at least 0.3 m from any side wall. If the pipe must be at a depth significantly less than this, it is recommended that it be positioned within 0.15 m of the side wall to reduce the possibility of vortex formation.

A faulty suction line is a source of many troubles in the proper working of pump. The suction pipe should be as short and direct as possible. Avoid a long circuitous path. If it is necessary to have a long suction line, the diameter of the pipe should be increased in order to reduce loss of head by friction. The diameter of the suction pipe should be one or two sizes more than the size of the flange. When an oversize pipe line is used, an eccentric reducer should be placed between it and the pump suction flange should be as shown in Fig. 3.5.

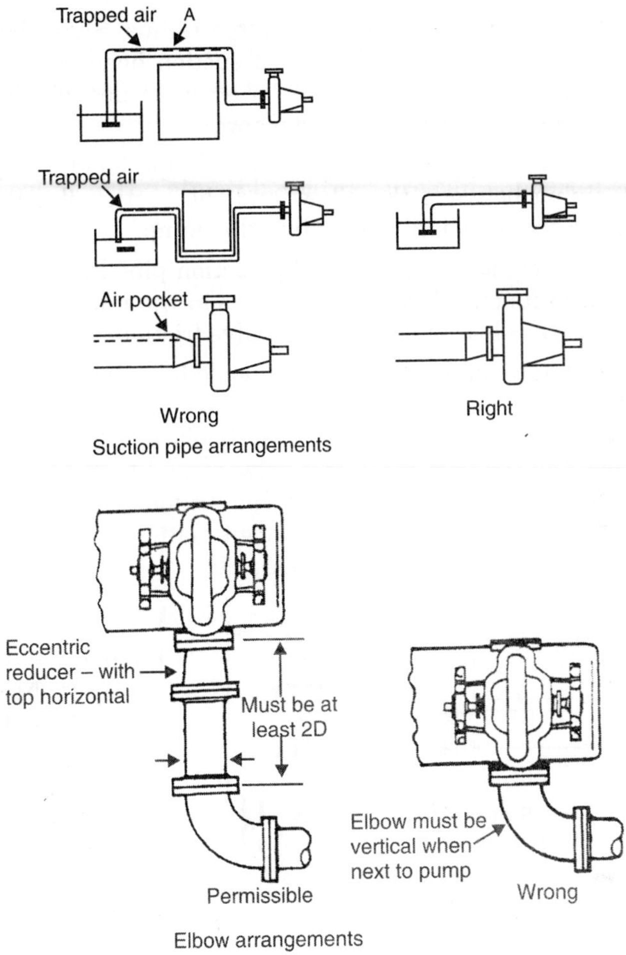

Fig. 3.5: Suction pipe connections to pump suction flange.

The suction pipe must be fitted vertically, so that water rises straight from the water source to the impeller of the pump. This may not be possible, all times, because the pump is generally installed on the bank of a well or on its side. One end of the suction pipe has to be immersed in water in the well at some distance away from the side of the well. Also you should see that no point along the suction pipe line is above the level of the pump (Fig. 3.5).

Long bends and other fittings next to the pump on the suction side should be selected and arranged carefully. If not the flow of water from the suction pipe line to the impeller will be adversely affected. Select long bends instead of elbows with a large radius. Sharp bends create more friction. These results in loss of suction head. So avoid elbows with sharp bends. In case, the use of elbow is unavoidable, insert a straight piece in between pump and elbow.

Provide 4 to 6 times the diameter of pipe length as possible of straight piping between the elbow and the suction flange. This will even out the flow of water as it enters the pump.

Do not provide any valve in the suction pipe line except the foot valve fitted to its end. If there are valves in the middle of the suction pipe line, there is loss of head due to valve friction.

The foot valve of the suction pipe should be well immersed in water. The depth of immersion should be more than 4 to 5 times the diameter of the suction pipe (Fig. 3.6). The foot valve should not also touch the bottom of the well. The strainer of the foot valve should have an area of 4 to 5 times the diameter of the suction pipe. The strainer will be necessary only, if the water is contaminated with foreign matter such as green plants growing in the water, fish, sand, pebbles, etc.

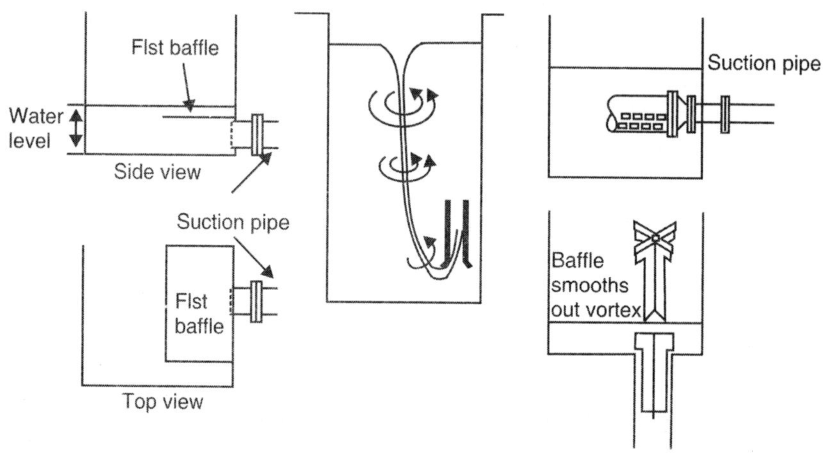

Fig. 3.6: Suction pipe arrangements.

As far as possible, see that the weight of the foot valve does not act on the suction pipe line itself in order to avoid distortions in alignment. One method will be to fix rails on the side of the well above the water level and hang the foot valve along with the suction pipe from the rails or give a cone shaped support from the suction well floor to the strainer bottom.

See that no water is discharged into the water source near the suction pipe. This may cause air to enter the suction pipe and cause loss of efficiency of the pump (Fig. 3.6).

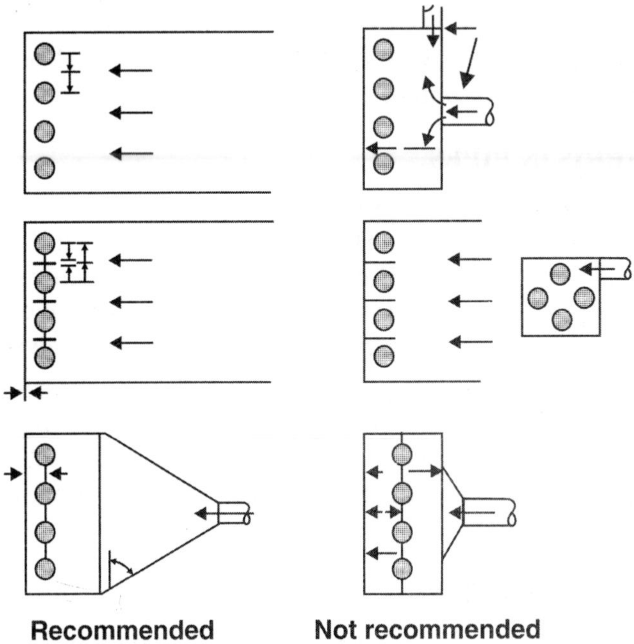

Fig. 3.7: Multiple pump arrangements in suction sump.

3.8.2 Delivery Pipe Work

It is essential to ensure that delivery pipe work is adequately supported and anchored to resist hydraulic thrust. When selecting valves and bends for delivery pipe work, the following points should be noted:

1. The diameter of the valve is not necessarily based on the diameter of the pipe work and is dependent on flow.
2. Non-return valves should be sized to give a maximum flow velocity of 3 m/s where a surge vessel is fitted to the pipeline or pumps operate in parallel, consideration should be given to fitting non-slam type non-return valves.
3. Where delivery pipes do not rise continuously and air pockets can form, it is important to provide some form of air release valve.

A non return valve and a gate valve are usually fitted to the delivery pipe. The gate valve is used to regulate the flow of water. The non return valve is a unidirectional valve and prevents back flow into the pump. The non return valve is placed between the gate valve and the pump, so that the pump may be inspected or repaired without emptying the delivery pipe. The gate valve is kept in a closed position while starting or stopping the pump, whenever any radial flow type centrifugal pumps are started. Where as valve should be in opened position for axial flow pumps.

3.9 DEBRIS IN PIPES

Another major source of trouble on site resulting from piping is the debris left inside the pipe work. It is important that all the pipe lines connecting to the pump on the suction side be thoroughly cleaned out before the pump is connected. Damage to pumps frequently occurs on commissioning due to nuts, bolts and tools being drawn through.

The effect of nuts, bolts, welding rods, etc. entering a pump designed to operate with tight running clearances for high efficiency can be imagined. This can be avoided by putting a mesh screen in to the pump suction temporarily during the initial running period upto the pipelines.

3.10 HYDRO–TESTING

Once the pipe work has been connected to the pump, it is advisable to hydro-testing the system. Leaks, particularly in the suction lines, can seriously reduce the pump's performance and indeed make it impossible to the pumps in order to start it.

3.11 ANCILLARY SUPPLIES

After the main pipe work and valves have been installed and checked, any ancillary supplies required by a pump are connected. The systems used depend on the particular application the pump, used. Typical examples of these services would be water flushing for the mechanical seal or stuffing box of the pump, water supply to cool the bearings or stuffing box and drain lines from pump or its base plate to carry away leakage and priming and venting pipe systems. Detailed requirements for any ancillary services must be provided by the pumps manufacturer prior to installation.

3.12 ELECTRICAL INSTALLATION

Once the mechanical work of installing the pump and driver set is completed, the electrical controls and instrumentation must be installed. The nature of this equipment varies considerably. This type of work is normally carried out by an electrical subcontractor, specializing in this type of work.

3.13 INITIAL START PREPARATIONS

3.13.1 DRIVE UNIT

The drive unit is checked to ensure that it is functioning correctly. In the case of an electric motor, it is essential to check the direction of rotation. This should be done with the shaft coupling disconnected as some pumps can be damaged by reverse rotation. The motor is kicked over and the rotation noted. If incorrect, two of the three electrical supply cables are interchanged to reverse the rotation.

3.13.2 ELECTRICAL CONTROLS

Control transducers are most commonly used such as, Pressure switches, flow switches, level probes, temperature measuring instrument, etc. Once the driver has been checked, the next step is to test the electrical controls and instrumentation. Five types of control transducers and checking is essential, so that the panel reacts correctly. Instrumentation usually has to be calibrated on detectors and limit switches. Controls are checked by carrying out trip tests; these consist of shorting each of the site. The procedures for this should be obtained from the instrument manufacturers. Where an electric motor is employed, the starter system is usually tested for correct operation together with any automatically operated valves or other equipment. It is most important to check the correct functioning of the electrical equipment before worst severe damage occur to the equipment.

3.14 ANCILLARY SERVICES

Any ancillary services required by the pump must now be tested. It is most important that a pump with an external flush supply to its packed gland or mechanical seal is not run without this supply, as severe over heating will occur after a short period of running with consequent damage occurring to the pump.

3.15 OPERATING PROCEDURE

The following procedure outlines the most important steps involved in pump operation. Any departure from this procedure must be based on good engineering practice.

3.15.1 PRIMING

Priming a centrifugal pump involves the removal of air, gas or vapour from the suction pipe work and pumps casing. As centrifugal pumps are generally not self-priming, they must be primed prior to start up.

If this is not done, the pump will not operate and, in addition, serious damage to water lubrication bushes could result.

The simplest method of priming is to arrange the pump so that it has a positive suction pressure. The pump can then be fitted by opening the air cock provided.

Radial centrifugal pumps often have rising power characteristics and are thus usually started against a valve position a positive suction is not possible, it is normal practice to fit a foot valve and strainer to the suction pipe inlet and use one of the following methods of priming:

1. Back priming from the delivery using a bypass round the no-return valve. It is important to ensure that the suction pipe work and valves are rated for the delivery pressure.
2. Using a tarnish fitted at the highest point on the pump casing, fill the pump and suction pipe work using an external water supply.
3. Vacuum priming equipment, such as an air or water operated ejector (jet pump) or a rotary exhauster, can be used to draw water from the suction in to the pump casing.

Venting is an operation carried out to remove any trapped air pockets in the system and may be necessary on the suction pipe line as well as the pump.

Many modern designs of pump are inherently self-venting. The volute of the pump is arranged so that air pockets cannot form. However, some types of pumps, do require venting. Normally suitable venting cocks or automatic vent valves are fitted. Pipe lines also often have air release valves fitted at high points to vent out trapped air pockets.

In a new system, the pump should stay fully primed from shut-down to start-up, marking further action unnecessary. However, foot valves are liable to wear or jam, and may lose the prime after a short period. It is advisable, therefore, to check the state of prime before start-up.

Caution: Such conditions, serious damage may result.

It is important that air does not leak into the pump during operation, as even if the pump during operates under

1. Ensure pump and suction pipe are correctly primed.
2. Ensure suction isolating valve (if fitted) is fully open.
3. Ensure delivery isolating valve is closed.

Note: Where starting conditions have been checked, the pump may be started with an open delivery valve.

4. Where a surge vessel is fitted in the pipeline, check that the air volume is correct.
5. Start the driving unit according to the manufacturer's instructions.
6. Open the pump delivery valve slowly until the required pressure or flow is obtained.

Caution:

(*a*) Prolonged running with a closed delivery valve can damage pump internal components and must be avoided.

(*b*) If the delivery pressure does not build up, stop the pump and repeat priming operation.

Note: Ideally, a pump should have a non-overloading power absorbed characteristic. The electrical equipment is then sized for this maximum power and the pump can then be operated over its whole range without danger of overloading.

However, in many cases, it is not possible to achieve this type of characteristic. Pumps of low specific speed (*i.e.*, centrifugal pumps) often have rising power characteristics and are thus usually started against a closed discharge valve. This reduces the total power demand required at start up and thus reduces the peak current drawn during the starting cycle. As soon as the pump is up to speed, the discharge valve is then opened fully. Pumps with a high specific speed, *i.e.* axial flow units, usually have a drooping power characteristic. These pumps are never started against a closed discharge valve; to do so would required a much larger motor than necessary for normal operation and additionally higher rated switchgear would be needed. This entails unnecessary cost. If it is essential for the system to have closed valve on start up, a bypass line can be installed.

3.16 STOPPING THE PUMPING EQUIPMENT

This procedure may seen very simple and in many cases it is. A typical sequence of operation for shutting down the pump set might be:

1. First shut down the driver (as the pump stops the discharge non-return valve shuts and prevents reverse flow of liquid through the pump).
2. If the pump is not required for some time, the suction and discharge isolating valves can be closed and any ancillary services such as cooling or flushing lines can be shut down.
3. Finally, in some instances, it may be desirable to drain the pump set, *i.e.* if maintenance is required.

This simple procedure, however can become rather more complicated. When pumps have to be stopped and started at frequent intervals, automatic control equipment is often used; this might consist of actuators for valves or systems to prevent the pump losing prime or system stop automatically re-prime the pump. Additionally, a fair amount of electrical control and instrumentation may also be used so that the pump sets can be monitored and operated from remote locations. Where an axial flow pump is used and operates against an open discharge, it may be found that, after switching the pump off, a siphon is created which will result in the reverse flow of water through the system. In this case, a siphon breaker system is obviously necessary.

3.17 ROUTINE RUNNING ATTENTION

As soon as the pump has run up to a speed, a number of routine checks must immediately be carried out.

1. Soft packed glands must be checked to ensure that there is a slight leakage from the gland. This lubricates and cools the gland. Excessively tight glands cause wear to the shaft or sleeve and the pump uses excessive power.
2. The bearings of the pump and driver must be checked to ensure that there is no overheating. As a general rule, it should be possible to hold a hand on a bearing housing without discomfort, although some designs do run at a greater temperature. A common cause of overheating with anti-friction bearings is over or under packing with grease.
3. Where auxiliary services are required by the pump, such as flushing or cooling supplies, these must be checked to ensure they are operating satisfactorily.

If possible, pressure gauge readings should be taken to ensure that the pump is operated at the correct duty. On the suction side, an unduly low reading may indicate a blockage in the line or a valve not fully open. Similarly on the discharge side, an unduly high reading would indicate the same. It is important that the pump runs within the correct band of differential head for which, it was designed for. When a differential pressure reading is too low, the pump may start to cavitate, which can cause damage if allowed to continue for a long period. A very low discharge pressure may be the result of a valve left open in the system, or the system characteristics not as correctly calculated.

It is equally important when monitoring the operation of a pump that the pump is not run for long periods at too high a differential pressure and hence at too low a flow rate. As a general indication, a pump should not be run at less than 10% of its best efficiency point flow rate. In many types of pump, operating at reduced flows can cause overheating in the pump. Since no pump is 100% efficiency, the inefficiency in the pump's conversion of power into hydraulic energy results in heat, which is transferred to the liquid passing through the pump.

The severity of the temperature rise in the pump depends on the ratio of the quantity of liquid passing through the pump to the amount of heat generated by the pump per unit time. When liquids containing abrasive particles are pumped, low flow rate operation also greatly increases the internal wear in the pump. The abrasive particles recirculate in the pump instead of passing straight through. If closed valve or low flow rate operation is unavoidable, a bypass system should be used and adjusted when the pump is running to maintain a minimum (about 15%) flow through the pump. Low flow operation also imposes unacceptably high thrusts on volute casing designs. This ultimately results in shaft failures.

Most pump installations with electric motor drivers have an ammeter installed. This instrument is a good indication of running problems. If the ammeter is running at a higher value than it should, then the pump may not be running at the correct duty. This can be checked with pressure gauges. Also, if the pump is not running freely, a packed gland may be too tight or solids in the liquid may have built up behind the impeller or in the eye ring areas.

Other problems to take care of, when the pump running are excessive vibration and noise from the pump. Vibration can be caused by misalignment of the pump and driver or distortion of the pump by the pipe work or distortion of the base plate due to foundations shifting. This can be checked and corrected if necessary. Vibration can also be caused by hydraulic problems, either in the system or in the pump. One of the examples of this is cavitation, which can occur under the conditions of low pressure of operation as well as high vacuum at suction. This problem will result in the pump running roughly and making loud cracking noises, which are caused by vapour bubbles collapsing in the pump water. Other restrictions in the system are: the use of short radius bends (elbows) at too high a flow rate and/or possible water hammer passages. Vibration can also be set up in the pipe work system. This may be due to resonance set up by throttling in the system on start up. All of these factors can cause a particular installation to vibrate or operate noisily. In a new system, the pump should stay fully primed from shut-down to start-up, marking further action unnecessary. However, foot valves are liable to wear or jam, and may lose the prime after a short period. It is advisable, therefore, to check the state of prime before start-up.

3.18 LUBRICATION

As with all other applicants, the same set of fundamentals apply to pump lubrication. The first fundamental aspects include:

- Correct viscosity grade or grease consistency.
- Compatibility of a lubricant with components, such as pump packings.
- External contributing factors, such as temperature extremes.

All these factors play a significant role in the efficient working function and life of a pump.

- Presence of water or chemicals.
- Length of service of the lubricant.
- Seal compatibility.
- External contributing factors, such as temperature extremes.

3.19 PUMP MAINTENANCE

Everyone who owns a pump would certainly like to know what steps should be taken to ensure that the pump gives long and trouble free service. Like any other machine, a pump too is a machine. Unless you take certain specific steps to maintain it properly your pump may develop trouble and cause you inconvenience.

To start with, we would like to say that the hints below apply mainly to an end-suction single stage centrifugal pump and not to other types of pumps.

Watch how your pump works from hour to hour. Hear what sound it makes. Is there a change in the sound? Every hour put your palm at the place where the pump has bearings to see if there is a sudden and unusual rise in temperature. If you observe any one of these, then you can be sure that the pump is developing trouble.

Observe also the water dripping from the portion where the pump's stuffing box is located. The amount of water inside the stuffing box should be just sufficient to cool and lubricate the coir packing and bush. If the dripping of water has ceased, then this is a sign of inadequate supply of water. On the other hand too much water should not also drip since that would be wasteful. If your pump is driven by an electric motor, please provide a good starter and switch which will protect your motor against overload etc.

Check the gland and the packings of the stuffing box once in six months. Verify if there is free movement. If there is no free movement, that should be set right. Change the packing at least once in 3 months. Clean the nuts and bolts of the above gland and oil them.

Check the grease on the ball bearings. If it is good enough there is no trouble. If the grease applied to the ball bearings is found to be melted or burnt, clean the ball bearings with kerosene oil. Then dry the ball bearings thoroughly. Apply fresh grease of correct grade. See that the quantity of grease applied is neither more nor less.

A thorough check of your pump once a year is necessary with the help of a mechanic. Remove the bearings and clean them with kerosene oil. Check for scratches on the balls, broken balls or other defects. After cleaning never place the ball bearings on mud or dust. Even after applying fresh grease, see that dust or dirt does not fall on the grease. Cover them with oil paper. Dust and dirt ruin the bearings. You should keep the bearings protected with oil paper till they are fitted into the pump.

Never dismantle a pump unless you have facilities to do so and unless you know the job. If you do not have the facilities or knowledge get the help of a mechanic.

If dismantling has to be done, proceed as follows:

Close the discharge valve and drain off the water in the pump casing. Remove the suction and delivery pipes from the pump casing. Remove the bolts fixing the motor to the base plate and put them in a small tin of kerosene oil. Now pull out the motor away from the pump. Then remove the bolts fixing the pump to the base plate. These bolts too should be put in kerosene oil. Remove the pump from the base plate.

Remove the half casing of the pump on the suction side by loosening the bolts which are used to connect the two half casings. Loosen the impeller nut and remove the impeller. NEVER use iron hammer or crow bars for removing use wooden hammers. Remove the bracket and casing if it is worn out. Otherwise do not remove it.

To remove the pump coupling loosen the coupling nut and remove the coupling. Now if you give a push from the impeller end of the shaft, the shaft will come out with the ball bearing in its seat. If the races are shaky, or if the balls have scratches or other flaws remove the ball bearing and replace a new one.

If the driving bushes of the flexible couplings are worn out, replace them with new ones.

Mark all the parts when you remove them. When reassembling follow the reverse order.

Remove the old packing and clean the stuffing box thoroughly with kerosene oil. Dry it. Wind the packing coir on the shaft of the pump and then cut the coir to form rings when the cut ends of each piece are put together. Insert each ring separately inside the stuffing box pushing it squarely into the stuffing box. See that it is seated firmly by pushing with the help of the gland itself. The joints of two consecutive rings should be about 120 degrees away from each other. In other words the joints of two successive coir rings should not be in one line, they should be staggered. After the required numbers of coir rings have been inserted tighten the gland nuts firmly. Make sure that the gland enters the stuffing box squarely and without twisting so that packing is subject to a uniform pressure on its entire periphery.

After tightening the gland loosen the nuts until they are just finger tight. When running in new packing it is advisable to allow leakage of more water at first from the stuffing box.

The stuffing box is so tightened that the leakage is sufficient to form a steady running stream rather than drops. With some experience, you will know what the optimum leakage should be.

Do not allow the leakage to fall over the bearing bed. If allowed, the bearing bed will get corroded and wear out. You should see that the leaking water is led away from the pump set through a small tube.

3.20 HOW BALL BEARINGS SHOULD BE TAKEN CARE OF?

Do not fill the bearing housing fully with grease. Fill up only one-third of the void space. This is because there must be enough gaps for the circulation of grease to dissipate the heat. Really more trouble is caused by excess of grease than the lack of it. Take care, that water does not enter the bearings. If water gets access, the bearings will fail.

Another enemy of the bearings is dirt and grit. They will damage the bearings. Grease serves as a good seal against dirt and grit. That is why grease is used for lubricating the ball bearings in pumps.

You may perhaps like to clean the ball bearings without removing them from the pump. Here is how you can do it. Remove the bearing cap and grease cup. Remove the hardened grease with a screw-driver or a similar tool. Inject into the bearing housing some kerosene oil, so that the grease thins out and runs out of the bearing housing. Then flush out with kerosene oil. Now add the correct amount of new grease and replace the grease cup with grease in it. See that the grease cup also is clean and that it contains fresh grease.

3.21 MAINTENANCE OF SHAFT, IMPELLER AND CASING

If properly maintained, a shaft will last as long as the pump lasts. When you overhaul a pump, check up the shaft, impeller and casing. Examine the shaft carefully. Sometimes the inner race of the ball bearings rotates on the shaft itself. This causes damage to the shaft. Check for the wear of the shaft due to packing rope. It is advisable to grind or polish with sand paper before assembling it. Use good packing coir.

A bent and distorted shaft is of no use. Never attempt to correct it. Better buy a new shaft.

If you find that the output of water given by the pump is getting reduced, then check the impeller. Remove the impeller from the casing of the pump examine carefully all its surfaces. If the surfaces show unusual wear replace a new impeller.

See that there is a close clearance between the casing and the impeller. If the clearance is more, consult a good mechanic as to how this may be rectified. When the pump is being over hauled replace the bush with a new one. Please quote the serial number of the pump and size when you order for a bush or other spare parts.

Always consult a reputed pump manufacturer, whenever you get doubts in the maintenance of pumps.

4 PUMPS: TROUBLES AND REMEDIES

The operation of a centrifugal pump may be affected by hydraulic or mechanical troubles. Hydraulic troubles may cause a pump to fail to deliver an insufficient capacity, develop insufficient pressure, lose its priming after starting or consume excessive power. Mechanical difficulties may appear at the stuffing boxes and bearings or produce vibration, noise or over-leaking of the pump.

In the table given below, the symptoms are grouped together, and the possible causes are numbered. The causes are given in a separate list and numbered. The remedies are also given in the tabulation, even though in many cases, it is self evident.

4.1 CENTRIFUGAL PUMP TROUBLES

(*a*) **Symptoms**	**Possible causes of troubles**
Pump does not deliver liquid	1, 2, 3, 4, 6, 11, 14, 16, 17, 22, 23
Insufficient capacity delivery	2, 3, 4, 5, 6, 7, 8, 9, 10, 11, 14, 17, 20, 22, 23, 29, 30, 31
Insufficient pressure developed	5, 14, 16, 17, 20, 22, 29, 30, 31
Pump loses prime after starting	2, 3, 5, 6, 7, 8, 11, 12, 13
Pump requires excessive power	15, 16, 17, 18, 19, 20, 23, 24, 26, 27, 29, 33, 34, 37
Stuffing box leaks excessively	13, 24, 26, 32, 33, 34, 35, 36, 38, 39, 40
Packing has short life	12, 13, 24, 26, 28, 32, 33, 34, 35, 36, 37, 38, 39, 40
Pump vibrates or is noisy	2, 3, 4, 9, 10, 11, 21, 23, 24, 25, 26, 27, 28, 30, 35, 36, 41, 42, 43, 44, 45, 46, 47

Bearing have short life	24, 26, 27, 28, 35, 36, 41, 42, 43, 44, 45, 46, 47
Pump overheats and seizes	1, 4, 21, 22, 24, 27, 28, 35, 36, 41

(*b*) Suction Troubles

(1) Pump not primed.

(2) Pump or suction pipe not completely filled with liquid.

(3) Suction lift too high.

(4) Insufficient margin between suction pressure and vapour pressure.

(5) Excessive amount of air or gas in liquid.

(6) Air pocket in suction line.

(7) Air leaks into suction line.

(8) Air leaks into pump through stuffing boxes.

(9) Foot valve too small.

(10) Foot valve partially clogged.

(11) Inlet of suction pipe insufficiently submerged.

(12) Water—Seal pipe plugged.

(13) Seal cage improperly located in stuffing box, preventing sealing fluid from entering space to form the seal.

(*c*) System Troubles

(14) Speed too low.

(15) Speed too high.

(16) Wrong direction of rotation.

(17) Total head of system higher than design head of pump.

(18) Total head of system lower than design head of pump.

(19) Specific gravity of liquid different from design.

(20) Viscosity of liquid differs from that for which it is designed.

(21) Operation at very low capacity.

(22) Parallel operation of pumps unsuitable for such operation.

(*d*) Mechanical Troubles

(23) Foreign matter in impeller.

(24) Misalignment.

(25) Foundations not rigid.

(26) Shaft bent.

(27) Rotating part rubbing on stationary part.

(28) Bearings worn.

(29) Wearing rings worn.

(30) Impeller damaged.

(31) Casing gasket defective, permitting internal leakage.

(32) Shaft or shaft sleeves worn or scored at the packing.

(33) Packing improperly installed.

(34) Incorrect type of packing for operating conditions.

(35) Shaft running off-center, because of worn out bearings or misalignment.

(36) Rotor out of balance, resulting in vibration.

(37) Gland too tight, resulting in, no flow of liquid to lubricate packing.

(38) Failure to provide cooling liquid to water cooled stuffing boxes.

(39) Excessive clearance at bottom of stuffing box between shaft and casing, causing packing to be forced into pump interior.

(40) Dirt or grit in sealing liquid, leading to scoring of shaft or shaft sleeve.

(41) Excessive thrust caused by a mechanical failure inside the pump or by the failures of the hydraulic balancing device, if any.

(42) Excessive grease or oil in antifriction bearing housing or lack of cooling, causing excessive bearing temperature.

(43) Lack of lubrication.

(44) Improper installation of antifriction bearings (damage during assembly, incorrect assembly or stacked bearings, use of unmatched bearings as a pair, etc.)

(45) Dirt getting into bearings.

(46) Rusting of bearings due to water getting into housing.

(47) Excessive cooling of water-cooled bearing, resulting in condensation in the bearing housing of moisture from the atmosphere.

Table 4.1: Pumps: Troubles and Remedies

Symptom	Possible causes	Remedy
Pump does not deliver water	1, 2, 3, 4, 6, 11, 14, 16, 17, 22, 23	Reprime the pump. Check whether the suction life as measured with a vacuum gauge does not exceed 7 metres. Reinstall the suction line such that it rises continuously towards the pump and eccentric reducers. Check whether the suction pipe is submerged sufficiently, i.e., at least 0.5 m below minimum water level. Check the motor speed and supply voltage. Change the motor connections. Check and if so, select another suitable pump for the operating head. Clean the impeller thoroughly and use. Check the air-rightness of suction pipe joints.
Insufficient capacity delivered	2, 3, 4, 5, 6, 7, 8, 9, 10, 11, 14, 17, 20, 22, 23, 29, 30, 31	Reprime the pump. Check and reinstall the pipe line such that it rises continuously towards the pump and use eccentric reducers. Tighten the packing. Replace the foot valve with one whose port area exceeds 2 times area of the pipe. Clean the foot valve thoroughly and use. The inlet of suction pipe should be at least 0.5. Meter below the minimum water level. Check the motor speed and line voltage. Check and select another suitable pump if necessary. Select another suitable pump designed to operate with the liquid to be handled. Clean the impeller thoroughly and use. Replace the worn wearing rings. Replace the impeller.

Insufficient pressure developed	5, 14, 16, 17, 20, 22, 29, 30, 31.	Check the motor speed and line voltage. Check with direction specified on pump and correct the line connection if necessary. Check and select another suitable pump if necessary. Replace the worn wearing rings. Select another suitable pump designed to operate with the liquid to be handled. Replace the impeller.
Pump loses prime after starting	2, 3, 5, 6, 7, 8, 11, 12, 13.	Reprime the pump. Reduce suction lift. Reinstall the suction pipe line such that it rises continuously towards the pump and use eccentric reducers. Blank off suction line and test with water pressure. Seal joints with washers and suitable compounds. The inlet of suction line should be at least 0.5 metres below minimum water level.
Pump requires excessive power	15, 16, 17, 18, 19, 20, 23, 24, 26, 27, 29, 33, 34, 37.	Reduce the speed to the rated value. Check with the direction specified on pump and change line connections if necessary. Check and select another suitable pump if necessary. Realign the pump correctly. Correct the shaft. Turn one of the part such that there is clearance between the two. Replace the worn wearing rings. Reinstall the packing properly. Loosen the gland until the water comes out drop by drop.
Stuffing box leaks excessively	13, 24, 26, 32, 33, 34, 35, 36, 38, 39, 40.	Realign the pump correctly. Correct the shaft. Replace the worn or scored parts. Rearrange the packing properly. Select and use best type of packing according to manufacturers' recommendations. Check and replace the worn bearings and realign the pump.

Packing has short life	12, 13, 24, 26, 28, 32, 33, 34, 35, 36, 37, 38, 39, 40.	Realign the pump properly. Correct the shaft. Replace the worn bearings. Replace the worn or scored parts. Select and use the best type of packing according to manufactures' recommendations. Check, correct the alignment and replace the worn bearings. Loosen the gland until water comes out drop by drop. Provide a suitable strainer in suction line to remove all dirt from the incoming liquid.
Pump vibrates or is noisy	2, 3, 4, 9, 10, 11, 21, 23, 24, 25, 26, 27, 28, 30, 35, 36, 41, 42, 43, 44, 45, 46, 47.	Reprime the pump. Suction life as measured with a vacuum. Gauge should not exceed 7 m of water check. Clean the foot valve thoroughly and use. The inlet of suction should be submerged at least 0.5 m below minimum water level. Do not operate it at very low capacity. Clean the impeller and use. Realign the pump properly. Put a rigid concrete foundation. Correct the shaft or replace. Replace the worn bearings. Replace the impeller. Replace the worn bearings and realign the pump properly. Balance the rotor properly and use. Check and put the correct amount of grease. Lubricate properly according to manufacturer's recommendations. Reinstall properly. Provide a water seal to the bearing housing.

Bearings have short life	24, 26, 27, 28, 35, 36, 41, 42, 43, 44, 45, 46, 47.	Realign the pump properly. Correct the shaft or replace if necessary. Turn one of parts providing a clearance between the parts. Replace the worn bearings. Replace the bearings and realign the pump. Balance the rotor thoroughly. Check and put in correct amount of grease. Lubricate properly according to manufacturer's recommendations. Reinstall correctly. Seal the bearing from dirt or water.
Pump overheats and seizes	1, 4, 21, 22, 24, 27, 28, 35, 36, 41.	Reprime the pump. Never operate it at very low capacity. Realign the pump properly. Replace the worn bearings. Replace the bearing and realign the pump. Balance the rotor properly.

4.2 PUMP FAULTS AND REMEDIES

Some faults that can be diagnosed by observation are listed in the following table:

Note: Should the remedy listed below entail working on the rotating components of the pump sets, the motors must be effectively isolated from the power supply to prevent accidental start-up during the remedial operation.

Table 4.2

S. no.	Symptom	Trouble fault	Remedy
1	Pump does not deliver liquid	Impeller rotating in wrong direction.	Reverse direction of rotation.
		Pump not properly primed Air or vapour lock in suction line.	Stop pump and reprime.
		Inlet of suction pipe insufficiently submerged.	Ensure adequate supply of liquid.
		Air leaks in suction line or gland arrangement.	Make good any leaks or repack gland.
		Pump not up to rated speed.	Increase speed.

2	Pump does not deliver rated quantity	Air or vapour lock in suction line.	Stop pump and reprime.
		Inlet of suction pipe insufficiently submerged.	Ensure adequate supply of liquid.
		Pump not up to rated speed.	Increase speed.
		Air leaks in suction line or gland arrangement.	Make good any leaks or repack gland.
		Foot valve or suction strainer chocked.	Clean foot valve or strainer.
		Restriction in delivery pipe work or pipe work incorrect.	Clear obstruction or rectify error in pipe work.
		Head underestimated.	Check head losses in delivery pipes, bends and valves reduce losses as required.
		Unobserved leak in delivery.	Examine pipe work and repair leak.
		Blockage in impeller or casing.	Remove half casing and clear obstruction.
		Excessive wear at neck rings or wearing plates.	Dismantle pump and restore clearances to original dimensions.
		Impeller damaged.	Dismantle pump and renew impeller.
		Pump gaskets leaking.	Renew defective gaskets.
3.	Pump does not generate its rated delivery pressure	Impeller rotating in wrong direction.	Reverse direction of rotation.
		Pump not up to rated speed.	Increase speed.
		Impeller neck rings worn excessive.	Dismantle pump and restore clearances to original dimensions.
		Impeller damaged or chocked.	Dismantle pump and renew impeller or clear blockage.
		Pump gaskets leaking.	Renew defective gaskets.

4.	Pump loses liquid after starting	Suction line not fully primed—air or vapour lock in suction line.	Stop pump and reprime.
		Inlet of suction pipe insufficiently submerged.	Ensure adequate supply of liquid at suction pipe inlet.
		Air leaks in suction line or gland arrangement.	Make good any leaks or renew gland packing.
		Liquid seal to gland arrangement logging ring (if fitted) chocked.	Clean out liquid seal supply.
		Logging ring not properly located.	Unpack gland and relocate logging ring under supply orifice.
5.	Pump overloads driving unit	Pump gaskets leaking.	Renew defective gaskets.
		Serious leak in delivery line, pump delivering more than its rated quantity.	Repair leaks.
		Speed too high.	Reduce speed.
		Impeller neck rings worn excessively.	Dismantle pump and restore clearances to original dimensions.
		Gland packing too tight.	Stop pump, close delivery valve to relieve internal pressure on packing, slacken back the gland nuts and retighten to finger tightness.
		Impeller damaged.	Dismantle pump and renew impeller.
		Mechanical tightness at pump internal components.	Dismantle pump, check internal clearances and adjust as necessary.
		Pipe work exerting strain on pump.	Disconnect pipework and realign to pump.

6.	Excessive vibration	Air or vapour lock in suction.	Stop pump and reprime.
		Inlet of suction pipe insufficiently submerged.	Ensure adequate supply of liquid at suction pipe inlet.
		Pump and driving unit incorrectly aligned.	Disconnect coupling and realign pump and driving unit.
		Worn or loose bearings.	Dismantle and renew bearings.
		Impeller choked or damaged.	Dismantle pump and clear or renew impeller.
		Rotating element shaft bent.	Dismantle pump and straighten or renew shaft.
		Foundation not rigid.	Remove pump, strengthen the foundation and reinstall pump.
		Coupling damaged.	Renew coupling.
		Pipe work exerting strain on pump.	Disconnect pipework and realign to pump.
7.	Bearing overheating	Pump and driving unit strain on pump.	Disconnect pipework and realign to pump and driving unit.
		Oil level too low or too high.	Replenish with correct grade of oil or drain down to correct level.
		Wrong grade of oil.	Drain out bearing flush through bearings; refill with correct grade of oil.
		Dirt in bearings.	Dismantle, clean out and flush through bearings; refill with correct grade of oil.
		Moisture in oil.	Drain out bearing, flush through and refill with correct grade of oil. Determine cause of contamination and rectify.
		Bearings too tight.	Ensure that bearings are correctly bedded to their journals with the correct amount of oil clearance Renew bearings if necessary.

		Too much grease in bearing.	Clean out old grease and repack with correct grade and amount of grease.
		Pipe work exerting strain on pump.	Disconnect pipework and realign to pump.
8.	Bearings wear	Pump and driving unit out of alignment.	Disconnect coupling and realign to pump.
		Rotating element shaft bent.	Dismantle pump, straighten or renew shaft. Renew bearings if necessary.
		Dirt in bearings.	Ensure that only clean oil is used to lubricate bearings. Renew bearings if necessary. Refill with clean oil.
		Lack of lubrication.	Ensure that only clean oil is used to lubricate clean oil.
		Bearing badly installed.	Ensure that bearings are correctly bedded to their journals with the correct amount of oil clearance. Renew bearings if necessary.
		Pipework exerting strain on pump.	Ensure that pipework is correctly aligned to pump. Renew bearings if necessary.
9.	Bearings wear	Excessive vibration.	Refer to Symptom 6.
10.	Irregular delivery	Air or vapour lock in suction line.	Stop pump and reprime.
		Fault in driving unit.	Examine driving unit and make good any defects.
		Air leaks in suction line or gland arrangement.	Make good any leaks or repack gland.
		Inlet of suction pipe insufficiently immersed.	Ensure adequate supply of liquid at suction pipe inlet.

		Air or vapour lock in suction line.	Stop pump and reprime.
		Inlet of suction pipe. Insufficiently submerged.	Ensure adequate supply of liquid at suction pipe inlet.
		Air leaks in suction line or gland arrangement.	Make good any leaks or repack gland.
11.	Excessive noise level	Pump and driving unit out of alignment.	Disconnect coupling and realign pump and driving unit.
		Worn or loose bearings.	Dismantle and renew bearings.
		Rotating element shaft bent.	Dismantle pump, straighten or renew shaft.
		Foundation not rigid.	Remove pump and driving unit, strengthen foundation.

4.3 MOTOR FAULTS AND REMEDIES

The following chart is restricted to general faults which would be investigated and rectified by a Pump service engineer, with assistance from licensed electrician.

Where faults of an electrical nature are suspected, which would require the machine to be isolated at the terminal box, a licensed electrician must be called in.

Table 4.3

Bearing overheating	Pump and driving unit out of alignment	Disconnect coupling and realign pump and driving unit.
	Oil level too low or too high	Replenish with correct grade of oil or drain down to correct level.
	Wrong grade of oil	Drain out bearing, flush through bearings; refill with correct grade of oil.
	Dirt in bearings	Dismantle, clean out and flush through bearings. Refill with correct grade of oil.
	Moisture in oil	Drain out bearing, flush through and refill with correct grade of oil. Determine cause of contamination and rectify.

	Bearings too tight	Ensure that bearings are correctly bedded to their journals with the correct amount of oil clearance. Renew bearings if necessary.
	Too much grease in bearing	Clean out old grease and repack with correct grade and amount of grease.
Bearings wear rapidly	Pump and driving unit out of alignment	Disconnect coupling and realign pump and driving unit. Renew bearings if necessary.
	Rotating element shaft bent	Dismantle motor, straighten or renew shaft. Renew bearings if necessary.
	Dirt in bearings	Ensure that only clean oil is used to replenish bearings. Renew bearings if necessary.
	Bearing badly installed	Ensure that bearings are correctly bedded to their journals with the correct amount of oil clearance. Renew bearings if necessary.
Excessive vibration	Pump and driving unit incorrectly aligned	Disconnect coupling and realign pump and driving unit.
	Worn or loose bearings	Dismantle and renew bearings.
	Rotating element shaft bent	Dismantle motor and straighten or renew shaft.
	Foundation not rigid	Remove motor strengthen the foundation and reinstall motor.
	Coupling damaged	Renew coupling.
Motor overheating	Motor incorrectly sized for duty	Check design requirements and replace motor if necessary.
	Cooling air ducts/fan chocked	Remove fan cowl and clean fan and ducting. On standard protected machines, it may be necessary to strip the motor and clean.
	Amps drawn too high due to under-voltage in supply	Check and rectify incoming supply.
	General ventilation in plant room inadequate	Investigate and improve ventilation.

Ammeter fluctuating during pump\ operation	Incoming electrical supply inconsistent	Contact supply authority to rectify supply.
	Loose connection in system	Check all connections in control panel, starters terminal boxes.
	Stator windings insulation breaking down	Remove cable from motor terminal box and mega test windings and cable separately. Replace cable or have motor windings rectified by rewinding, reinsulation, backing.
	On slip ring motor brushes not contacting slip rings correctly	Expose slip rings and check and rectify brushes, brush holders, pressure springs.
	Ammeter faulty	Check and replace if necessary.
Motor will not start	Supply voltage not reaching motor	Investigate and rectify.
	Connections in system faulty or cable broken	Tighten and clean all connections and check cable for continuity.
	Motor incorrectly connected	Check connections against diagram.
	Motor incorrectly sized for application	Try starting motor uncoupled. Check sizing of motor to requirements of pump. If necessary change to motor more suited to requirements i.e., high starting torque, etc.
	Starter contactor faulty	Check and rectify.
	Auto transformer faulty	Check and rectify.
	Brushes not contacting slip rings correctly	Expose slip rings and check and rectify brushes, brush holders and pressure springs.
	Internal connections in motor faulty	Check and rectify.
	Rotor resistance incorrect	Check design and rectify rotor.
	Overloads in contactor incorrectly adjusted	Check and reset to required value.

Motor starts but will not take load	Overloads in contactor incorrectly adjusted	Check and reset to required value.
	Supply voltage incorrect	Check and rectify.
	Loose connections in system	Check and rectify.
	Motor incorrectly sized for duty	Check design and rectify.
	Method of starting incorrect for duty requirement	Check design and rectify.
	Motor design incorrect	Check design and rectify.
	Driven unit excessively stiff	Check and rectify.
	Direction of rotation incorrect	Check and rectify.
Brushes on slip ring motor spark continuously	Motor being overloaded	Check design and rectify.
	Brushes installed are the wrong grade	Check with original supplier/ manufacturer.
	Brushes sticking in holders	Clean and refit brushes and check pressure springs.
	Brushes need bedding	See supplier/manufactures handbook.
	Brush pressure incorrectly adjusted	See supplier/ manufacturers handbook.
	Slip rings dirty/worn	Check and clean, restore to original design condition replace brushes.
	Machine vibrating excessively due to poor bearings, alignment, foundations	Replace bearings, check alignment, repair/rebuilt foundation.
	Moisture enters windings	Find probable point of entry and plug, have stator rewound.
	Condensation forming due to anti condensation heaters not working	Check a/c heaters for operation and sizing.

Motor stator windings burnt out	Overloads set too high and/or motor too small for job	Check overload settings and design for sizing motor.
	Design of motor incorrect for application, i.e., SPDP instead of TEFC, thus allowing moisture entrance	Check suitability of design to actual installed item.
	Class of insulation not suited to application, i.e., class E instead of B or F, etc.	Check suitability of design to actual installed item.

INDEX